Translations of
MATHEMATICAL MONOGRAPHS

Volume 36

Theory of Convex Programming

E. G. Gol′šteĭn

American Mathematical Society
Providence, Rhode Island

ВЫПУКЛОЕ ПРОГРАММИРОВАНИЕ
ЭЛЕМЕНТЫ ТЕОРИИ

Е. Г. ГОЛЬШТЕЙН

Издательство "Наука"
Главная редакция
Физико-Математической Литературы
Москва 1970

Translated from the Russian by K. Makowski

2000 *Mathematics Subject Classification.* Primary 49A50, 90C25;
Secondary 52A40, 90D05.

Library of Congress Cataloging-in-Publication Data

Gol′shteĭn, Evgeniĭ Grigor′evich.
Theory of convex programming.
(Translations of mathematical monographs, v. 36)
Translation of Vypukloe programmirovanie.
1. Convex programming. I. Title. II. Series.

T57.4.G6413 519.7′6 72-3180
ISBN 0-8218-1586-5

AMS softcover ISBN: 978-0-8218-4623-0

∞ The paper used in this book is acid-free and falls within the guidelines
established to ensure permanence and durability.
Visit the AMS home page at http://www.ams.org/

10 9 8 7 6 5 4 3 2 1   13 12 11 10 09 08

FOREWORD

The lectures on convex programming theory which the author presented in the summer of 1968 to the students of the All-Union School on Mathematical Programming in the city of Alma-Ata were used as the basis for this book.

It is well known that duality theory and the corresponding dual approach to the analysis of linear problems play a significant role in linear programming. On the one hand, this approach allows one to give sufficiently precise mathematical descriptions of a number of economic mechanisms. On the other hand, it leads toward effective computational methods of linear programming. A significant place in the book is given to the construction of a duality theory for convex programming problems. A pair of dual convex programming problems is closely connected with the problem of finding a saddle point of a Lagrangian function. Therefore, the approach to the analysis of convex programming problems which we develop here is closely related to the well-known approach of Kuhn and Tucker [**5**], [**11**]. However, as we shall show, the manner of presentation adopted in the book allows us to construct a more general theory of convex programming.

The method of constructing a theory of dual convex programming problems adopted in the book is not entirely traditional (see [**1**] and [**12**]). Here, the basic theorem on antagonistic games due to John von Neumann was taken as the starting point (usually, the connection between convex programming theorems and von Neumann's theorem is not noted).

Besides the basic facts of convex programming theory, the book includes a number of results on the marginal values of convex programming problems.

Although the presentation is carried out for the case of finite-dimensional problems, many of the theorems remain in force for more general infinite-dimensional problems. Moreover, the transition to functional spaces does not, as a rule, involve any change in the structure of the corresponding arguments.

The author

Table of Contents

§1. The dual problem

By a general convex programming problem one understands the problem of finding the supremum of a function $f(x)$

$$(1) \qquad f(x) \rightarrow \sup$$

subject to the constraints

$$(2) \qquad f_i(x) \geqq 0, \qquad i = 1, 2, \cdots, m;$$

$$(3) \qquad x \in G,$$

where $x = (x_1, x_2, \cdots, x_n)$, f and f_i are concave functions defined on G, and G is a convex subset of an n-dimensional space.

We remind the reader that a set G is said to be *convex* if, for any points $x', x'' \in G$ and an arbitrary number λ, $0 < \lambda < 1$,

$$x_\lambda = \lambda x' + (1 - \lambda) x'' \in G.$$

We also recall that a function $f(x)$ defined on a convex set G is said to be *concave* (*convex*) if, for any two points x', $x'' \in G$ and an arbitrary number λ, $0 < \lambda < 1$,

$$f(\lambda x' + (1 - \lambda) x'') \geqq (\leqq) \lambda f(x') + (1 - \lambda) f(x'').$$

We denote by R the set of all points which satisfy constraints (2) and (3):

$$R = \{x : f_i(x) \geqq 0, \ i = 1, 2, \cdots, m; \ x \in G\}.$$

It follows immediately from the fact that G is convex and f and f_i are concave that the set R is convex. Therefore, problem (1) – (3) consists of maximizing a concave function $f(x)$ on the convex set R. Formally, one can write down the constraints of problem (1) – (3) as follows: $x \in R$. However, this form does not make it possible to study in detail each of the constraints (2) which take part in forming R. In the sequel, we shall assign a variable of the dual problem to each of constraints (2). For this reason, a representation of R in the form (2) and

(3) will be more convenient. We note that such a representation can be made in more than one way. Usually, one collects in the set G the points which satisfy simple constraints of the forms $x \geqq 0$, $\alpha \leqq x \leqq \beta$, etc. However, in a number of cases (e.g., in the theory of block programming), it turns out to be useful to consider more complicated sets G. In what follows, we shall mean by G an arbitrary convex set.

We introduce several definitions. We shall call the points $x \in R$ *feasible solutions* of problem (1) – (3). A sequence $X = \{x^{(k)}\}$ of feasible solutions $x^{(k)}$ will be called an *optimizing sequence* (*of solutions*), if $\lim_{k\to\infty} f(x^{(k)})$ exists (finite or infinite). We set

$$f(X) = \lim_{k\to\infty} f(x^{(k)}).$$

An optimizing sequence of solutions X^* is said to be a *solution* of the problem, if

$$f(X^*) = \sup_{x \in R} f(x).$$

Any problem of the (1) – (3) type has some optimizing sequence as a solution, but it is easy to construct examples of problems which do not have feasible solutions as solutions. Exactly this fact has forced us to broaden the set of feasible solutions by introducing the notion of an optimizing sequence.

We form the so-called Lagrangian function for problem (1) – (3):

$$F(x,y) = f(x) + \sum_{i=1}^{m} y_i f_i(x), \quad x \in G, \quad y \geqq 0,$$

where $y = (y_1, y_2, \cdots, y_m)$. With the aid of this function, we shall represent our problem in an equivalent form which is more convenient for the subsequent analysis.

We set

$$\varphi(x) = \inf_{y \geqq 0} F(x,y), \quad x \in G, \tag{4}$$

and consider the problem of maximizing the function $\varphi(x)$ on the set G:

$$\varphi(x) \to \sup, \tag{5}$$

$$x \in G. \tag{6}$$

We set

$$v = \begin{cases} \sup_{x\in R} f(x), & \text{if } R \neq \emptyset, \\ -\infty, & \text{if } R = \emptyset; \end{cases}$$

$$v' = \sup_{x\in G} \varphi(x).$$

LEMMA 1. *Problems* (1) – (3) *and* (4) – (6) *are equivalent, i.e.,*

(i) $v = v'$;

(ii) *if* $X = \{x^{(k)}\}$ *is a solution of problem* (1) – (3), *then it is also a solution of problem* (4) – (6);

(iii) *if* $X = \{x^{(k)}\}$ *is a solution of problem* (4) – (6) *with*

$$\varphi(X) = \lim_{k\to\infty} \varphi(x^{(k)}) > -\infty,$$

then X *is a solution of problem* (1) – (3).

PROOF. Let $x \in R$. Since $f_i(x) \geqq 0$, $i = 1, 2, \cdots, m$, we have, for any $y \geqq 0$,

$$F(x, y) \geqq f(x).$$

On the other hand, $F(x, 0) = f(x)$. Therefore,

$$\varphi(x) = \inf_{y\geqq 0} F(x, y) = F(x, 0) = f(x).$$

If $x \notin R$, then $f_i(x) < 0$ for at least one value of the index i. Let, e.g., $f_1(x) < 0$. We set $y^{(k)} = (k, 0, 0, \cdots, 0)$. Obviously,

$$\varphi(x) = \inf_{y\geqq 0} F(x, y) \leqq \lim_{k\to\infty} F(x, y^{(k)}) = -\infty.$$

Thus,

$$\varphi(x) = \begin{cases} f(x) & \text{for } x \in R, \\ -\infty & \text{for } x \notin R. \end{cases} \tag{7}$$

If $v = -\infty$, i.e., if R is an empty set, then by (7),

$$\varphi(x) \equiv -\infty, \quad \text{i.e.,} \quad v' = -\infty.$$

If $v > -\infty$, i.e., $R \neq \emptyset$, then by (7),

$$v' = \sup_G \varphi(x) = \sup_{x\in R} f(x) = v.$$

Assertion (i) has been proved.

Assume that $X = \{x^{(k)}\}$ is a solution of problem (1) – (3), i.e.,

$$\lim_{k\to\infty} f(x^{(k)}) = v > -\infty.$$

Since $x^{(k)} \in R$, we have by (7),

$$v = \lim_{k\to\infty} f(x^{(k)}) = \lim_{k\to\infty} \varphi(x^{(k)}).$$

Further, we obtain, taking (i) into account,

$$\lim_{k\to\infty} \varphi(x^{(k)}) = v',$$

which proves assertion (ii). A proof of assertion (iii) of the lemma is similar.

We note that, in our proof of Lemma 1, we did not make use of the fact that (1) – (3) is a convex programming problem. Problems (1) – (3) and (4) – (6) are equivalent for any functions f and f_i and any set G. The lemma shows that the initial problem (1) – (3) is equivalent to the problem of finding

$$\sup_{x\in G} \inf_{y\geqq 0} F(x,y), \tag{8}$$

where the operations "inf" and "sup" are taken in exactly the order fixed in relation (8).

In order to formulate the problem dual to (1) – (3), we interchange the places of "inf" and "sup" in formula (8). We set

$$\psi(y) = \sup_{x\in G} F(x,y). \tag{9}$$

The problem dual to (1) – (3) consists, by definition, of minimizing the function $\psi(y)$ on the set $\{y \geqq 0\}$:

$$\psi(y) \to \inf, \tag{10}$$

$$y \geqq 0. \tag{11}$$

We denote by v the extremal value in the dual problem, i.e., the infimum of $\psi(y)$ on the positive orthant $y \geqq 0$.

We shall illustrate the above general scheme for forming a dual problem on an example of a general linear programming problem. We consider the following linear programming problem:

$$\sum_{j=1}^{n} c_j x_j \to \max, \tag{12}$$

$$\sum_{j=1}^{n} a_{ij} x_j \leqq b_i, \qquad i = 1, 2, \cdots, m, \tag{13}$$

$$x_j \geqq 0, \qquad j = 1, 2, \cdots, n. \tag{14}$$

In this case,

$$f(x) = \sum_{j=1}^{n} c_j x_j, \quad f_i(x) = b_i - \sum_{j=1}^{n} a_{ij} x_j, \quad i = 1, 2, \cdots, m;$$

$$G = \{x : x \geqq 0\}.$$

Therefore

$$\begin{aligned} F(x, y) &= \sum_{j=1}^{n} c_j x_j + \sum_{i=1}^{m} y_i \left(b_i - \sum_{j=1}^{n} a_{ij} x_j \right) \\ &= \sum_{j=1}^{n} c_j x_j + \sum_{i=1}^{m} b_i y_i - \sum_{i=1}^{m} \sum_{j=1}^{n} a_{ij} y_i x_j. \end{aligned}$$

In accordance with the general rule of forming a dual problem, we introduce the function

$$\begin{aligned} \psi(y) &= \sup_{x \in G} F(x, y) \\ &= \sup_{x \geqq 0} \left[\sum_{j=1}^{n} c_j x_j + \sum_{i=1}^{m} b_i y_i - \sum_{i=1}^{m} \sum_{j=1}^{n} a_{ij} y_i x_j \right] \\ &= \sum_{i=1}^{m} b_i y_i + \sup_{x \geqq 0} \sum_{j=1}^{n} \left(c_j - \sum_{i=1}^{m} a_{ij} y_i \right) x_j. \end{aligned}$$

We set

$$c_j(y) = c_j - \sum_{i=1}^{m} a_{ij} y_i, \qquad j = 1, 2, \cdots, n.$$

Obviously,

$$\sup_{x \geqq 0} \sum_{j=1}^{n} c_j(y) x_j = \sum_{j=1}^{n} \sup_{x_j \geqq 0} [c_j(y) x_j].$$

Further

$$\sup_{x_j \geqq 0} [c_j(y) x_j] = \begin{cases} 0 & \text{for } c_j(y) \leqq 0, \\ \infty & \text{for } c_j(y) > 0. \end{cases}$$

Therefore, if y satisfies the constraints $c_j(y) \leqq 0, j = 1, 2, \cdots, n$, then

$$\sup_{x \geqq 0} \sum_{j=1}^{n} c_j(y) x_j = 0 \quad \text{and} \quad \psi(y) = \sum_{i=1}^{m} b_i y_i.$$

In the opposite case

$$\sup_{x \geqq 0} \sum_{j=1}^{n} c_j(y) x_j = \infty \quad \text{and} \quad \psi(y) = \infty.$$

Let $T = \{ y : c_j(y) \leqq 0,\ j = 1, 2, \cdots, n \}$. We have shown that

$$\psi(y) = \begin{cases} \sum_{i=1}^{m} b_i y_i & \text{for } y \in T, \\ \infty & \text{for } y \notin T. \end{cases}$$

Since the dual problem consists of minimizing the function $\psi(y)$, it is sufficient to investigate this function only at the points y for which $\psi(y) < \infty$. In other words, the problem dual to (12) – (14) consists of minimizing $\psi(y)$ on the set of nonnegative $y \in T$. Thus the final formulation of the dual problem has the form

$$\sum_{i=1}^{m} b_i y_i \to \min, \tag{15}$$

$$c_j(y) = c_j - \sum_{i=1}^{m} a_{ij} y_i \leqq 0, \qquad j = 1, 2, \cdots, n, \tag{16}$$

$$y_i \geqq 0, \qquad i = 1, 2, \cdots, m. \tag{17}$$

We have obtained a well-known form of the dual linear programming problem.

We remind the reader that the constraints of this problem delimit the set of the points y at which $\psi(y) < \infty$. Similarly, the general scheme for constructing dual problems can be particularized also to various other special classes of convex programming problems (e.g., to quadratic programming problems). As in the case which we have just considered, the additional constraints of the dual problem arise as a result of eliminating the points at which $\psi(y) = \infty$. However, we shall not dwell on this any longer.

§2. The basic theorem on antagonistic games and its generalizations

By a duality theorem we shall understand a result which, under certain assumptions, guarantees the following equality, usually called the duality relation:

$$v = \tilde{v}.$$

Here, by Lemma 1 and the notation adopted earlier,

$$v = \begin{Bmatrix} \sup_{x\in R} f(x), & R \neq \emptyset \\ -\infty, & R = \emptyset \end{Bmatrix} = \sup_{x\in G} \varphi(x) = \sup_{x\in G} \inf_{y\geqq 0} F(x,y);$$

$$\tilde{v} = \inf_{y\geqq 0} \psi(y) = \inf_{y\geqq 0} \sup_{x\in G} F(x,y).$$

Thus the duality relation is equivalent to the equality

$$(18) \qquad \sup_{x\in G} \inf_{y\geqq 0} F(x,y) = \inf_{y\geqq 0} \sup_{x\in G} F(x,y),$$

where $F(x,y) = f(x) + \sum_{i=1}^{n} y_i f_i(x)$ is the Lagrangian function for problem (1) – (3). Since f and f_i are concave and $y \geqq 0$, the function $F(x,y)$ is concave in x. Moreover, it is obviously linear in y. Therefore, if the convex sets on which the supremum and infimum are taken in (18) were compact, equality (18) would follow from von Neumann's theorem on antagonistic games [7]. However, in our case one of these sets is certainly unbounded, and generally speaking the second one (G) is also unbounded. This explains the fact that the duality relations for convex programming problems hold only if some additional assumptions on the problems under investigation are satisfied. This also seems to be the reason why von Neumann's theorem has not been applied, until now, in convex programming theory: in order to obtain various duality theorems, one usually employs special constructions and considerations. Below we shall attempt to abandon the traditional way of presenting the convex programming theory. As far as possible, we shall base everything on von Neumann's theorem or, more precisely, on one of its extensions. In the first place, such an approach makes clear the natural connection between the theory of antagonistic games and convex programming and, secondly, it allows us to obtain more general results.

Let us forget, for the time being, our problem (1) – (3), which gave rise to the function $F(x,y)$, and let us consider a function $f(x,y)$ of an n-dimensional vector $x \in X$ and an m-dimensional vector $y \in Y$. The sets X and Y on whose product $f(x,y)$ is defined are assumed to be convex. The function $f(x,y)$ is concave in $x \in X$ for any fixed $y \in Y$ and convex on Y for any fixed $x \in X$. We shall rely on the following version of von Neumann's theorem, which is due to H. Kneser.

THEOREM 1. *If one of the sets X and Y is compact (i.e., bounded and closed) and if the function $f(x,y)$ is continuous in the corresponding variable, then*

$$\sup_X \inf_Y f(x,y) = \inf_Y \sup_X f(x,y). \tag{19}$$

First of all, we shall prove a very simple assertion.

LEMMA 2. *Regardless of the nature of the sets X and Y and the function $f(x,y)$,*

$$\sup_X \inf_Y f(x,y) \leqq \inf_Y \sup_X f(x,y). \tag{20}$$

PROOF. Obviously, for any $x \in X$ and $y \in Y$,

$$\varphi(x) = \inf_Y f(x,y) \leqq \sup_X f(x,y) = \psi(y).$$

Taking the supremum over x in the left-hand side of this inequality and the infimum over y in the right-hand side, we obtain (20).

The proof of Theorem 1 rests on two well-known theorems of functional analysis. We think it is worthwhile to formulate them here. Let B be a Banach space, and let K be a convex subset of B. The following assertion is known as a *separation theorem*.

If the convex set K contains interior points, and if a point $p \in B$ is not an interior point of K, then there exists a nonzero, bounded linear functional $l(z)$ defined on B such that $l(z) \leqq l(p)$ for every element $z \in K$.

Let X be a compact set in a finite-dimensional space. We consider the Banach space $C(X)$ of all real-valued functions continuous on X with the norm of an element $\varphi(x)$ given by the relation

$$|\varphi| = \max_{x \in X} |\varphi(x)|.$$

In the proof of Theorem 1, we shall use the separation theorem for a subset of $C(X)$. For this reason, we shall need to know the general form of a bounded linear functional defined on $C(X)$. We introduce a collection S of subsets σ of X by means of the following requirements:

(i) S contains all the open subsets of X and, for any $\sigma \in S$, we have $\bar{\sigma} = X \setminus \sigma \in S$;

(ii) together with any finite or countable sequence of its elements, S contains their union;

(iii) S is the smallest set which has properties (i) and (ii).

The set S thus introduced is said to be the *collection of Borel subsets* of X.

Let μ be a real-valued function defined on the set S of all Borel subsets of X. The function μ is said to be a *measure* on X if, for every (countable) sequence of nonintersecting subsets $\sigma_i \in S$, $i = 1, 2, \cdots$,

$$\mu\left(\bigcup_i \sigma_i\right) = \sum_i \mu(\sigma_i).$$

The total variation of μ on X is the number

$$\operatorname{var} \mu_X = \sup \sum_i |\mu(\sigma_i)|,$$

where the supremum is taken over all sequences which consist of nonintersecting sets $\sigma_i \in S$ (and which are countable). If $\operatorname{var} \mu_X < \infty$, then the measure μ is said to be *bounded*.

Let μ be an arbitrary bounded measure on X. One can introduce, in the usual manner, the notion of a function which is measurable with respect to μ, and for any bounded function $\varphi(x)$, $x \in X$, measurable with respect to μ, one can define the integral with respect to the measure μ on the set X: $\int_X \varphi(x)\, d\mu$ (this is done in exactly the same way as when μ is the usual Lebesgue measure). In particular, any continuous function on X turns out to be integrable. We shall use the following theorem of F. Riesz:

If $l(\varphi)$ is a bounded linear functional defined on $C(X)$, then there exists a bounded measure μ such that

$$l(\varphi) = \int_X \varphi(x)\, d\mu, \quad |l| = \operatorname{var} \mu_X.$$

Moreover, if l is nonnegative, i.e., if $l(\varphi) \geqq 0$ for any function $\varphi(x)$ which is nonnegative on X, then μ is nonnegative and therefore $|l| = \mu(X)$.

Let $\varphi(x)$ be a continuous function, defined and concave on a convex compact set X, and let μ be a nonnegative measure on X with $\mu(X) = 1$. We set $x_\mu = \int_X x\, d\mu$. We shall show that

$$\varphi(x_\mu) \geqq \int_X \varphi(x)\, d\mu, \quad x_\mu \in X. \tag{21}$$

Indeed, the integrals in (21) can be replaced, with arbitrary accuracy, by the corresponding integral sums, i.e., for any $\epsilon > 0$ there exist $x_i \in X$ and $\mu_i \geqq 0$ such that

$$\left| x_\mu - \sum_i \mu_i x_i \right| \leqq \epsilon, \quad \left| \int_X \varphi(x)\, d\mu - \sum_i \varphi(x_i) \mu_i \right| \leqq \epsilon, \tag{22}$$
$$\sum_i \mu_i = 1.$$

Since X is convex and $\varphi(x)$ is concave,

$$\varphi\left(\sum_i \mu_i x_i\right) \geqq \sum_i \mu_i \varphi(x_i), \quad \sum_i \mu_i x_i \in X.$$

The desired relations (21) are a consequence of these last inequalities, inequality (22), and the fact that $\epsilon > 0$ is arbitrary. After these preliminary remarks, we shall now prove Theorem 1.

PROOF OF THEOREM 1. For the sake of definiteness, let X be compact and let $f(x, y)$ be continuous in x for any fixed $y \in Y$. We shall consider the space $C(X)$ of all continuous functions on the compact set X. We define the set $K \subset C(X)$ by the following condition:

$$K = \{\varphi(x) : \varphi(x) \geqq f(x, y)\}$$

for $y \in Y$ and every $x \in X$. The set K is convex. Indeed, if

$\varphi_1(x)$ and $\varphi_2(x)$ are arbitrary elements of K and $0 < \alpha < 1$, then

$$\varphi_\alpha(x) = \alpha\varphi_1(x) + (1-\alpha)\varphi_2(x)$$
$$\geqq \alpha f(x,y_1) + (1-\alpha)f(x,y_2), \quad y_1, y_2 \in Y.$$

Since $f(x,y)$ is convex in y and Y is convex,

$$\alpha f(x,y_1) + (1-\alpha)f(x,y_2) \geqq f(x,y_\alpha),$$

where $y_\alpha = \alpha y_1 + (1-\alpha)y_2 \in Y$.

Thus we have found a point $y_\alpha \in Y$ such that

$$\varphi_\alpha(x) \geqq f(x,y_\alpha), \qquad x \in X,$$

i.e., $\varphi_\alpha(x) \in K$. The convexity of K has been proved. The set K contains interior points. Indeed, let $y \in Y$. Then $\bar{\varphi}(x) = f(x,y) + 1$ is an interior point of K since, for any element $\varphi(x) \in C(X)$ with norm not greater than one,

$$\bar{\varphi}(x) + \varphi(x) \geqq f(x,y), \qquad x \in X,$$

i.e., $\bar{\varphi}(x) + \varphi(x) \in K$.

Thus K is a convex set which contains interior points. Let $\varphi_v^*(x) \equiv v$, $x \in X$. We set

$$v_1 = \sup_X \inf_Y f(x,y), \qquad v_2 = \inf_Y \sup_X f(x,y).$$

Since $\sup_X f(x,y) < \infty$ for any $y \in Y$, $v_2 < \infty$. The rest of the argument is carried out under the assumption that $v_2 > -\infty$. Let us verify that, for any $\epsilon > 0$,

$$\varphi^*_{v_2-\epsilon}(x) \notin K. \tag{23}$$

Indeed, if we assume the contrary, then there exists a point $y_\epsilon \in Y$ such that $v_2 - \epsilon \geqq f(x,y_\epsilon)$, $x \in X$. We have from this relation

$$v_2 - \epsilon \geqq \sup_X f(x,y_\epsilon) \geqq \inf_Y \sup_X f(x,y) = v_2, \quad \text{i.e.,}\ \epsilon \leqq 0.$$

But $\epsilon > 0$ by the assumption. This contradiction proves (23). Relation (23), which holds for any $\epsilon > 0$, means that the element $\varphi^*_{v_2}(x)$ is not an interior point of K.

We now apply the separation theorem to the set K and the point $\varphi^*_{v_2} = \varphi^*_{v_2}(x)$. According to this theorem, there exists a nonzero bounded linear functional $l(\varphi)$, defined on $C(X)$

and such that

$$l(\varphi) \geqq l(\varphi^*_{v_2}) \tag{24}$$

for every point $\varphi \in K$. We shall prove that l is nonnegative. For this purpose, we assume that there exists a nonnegative continuous function $\varphi_0(x) = \varphi_0 \in C(X)$ for which

$$l(\varphi_0) < 0. \tag{25}$$

Without loss of generality, we may assume that

$$\varphi_0(x) > \delta > 0, \qquad x \in X, \tag{26}$$

since otherwise we would replace $\varphi_0(x)$ by the function $\varphi_0(x) + \varphi_\delta^*(x)$ which, for δ sufficiently small, satisfies (25) because l is bounded. It follows from inequality (26) and from the definition of K that, for $\lambda > 0$ sufficiently large,

$$\lambda\varphi_0(x) \in K.$$

Hence, taking into account (24), we have

$$l(\varphi_0) \geqq \lim_{\lambda \to \infty} \left[\frac{1}{\lambda} l(\varphi_{v_2}) \right] = 0,$$

contrary to assumption (25). This contradiction proves that l is nonnegative.

We shall now use the theorem of F. Riesz, according to which a nonnegative bounded linear functional can be represented in the form

$$l(\varphi) = \int_X \varphi(x)\, d\lambda, \tag{27}$$

where λ is a nonnegative measure on the compact set X, and $|l| = \lambda(X) > 0$. We rewrite (24), taking into account representation (27):

$$\int_X \varphi(x)\, d\lambda \geqq \int_X v_2 d\lambda = v_2 \cdot \lambda(X) \tag{28}$$

for any function $\varphi(x) \in K$. If we now set $\lambda^* = \lambda/\lambda(X)$ and $\varphi(x) = f(x,y) \in K$ for any fixed $y \in Y$, then we obtain, by (28),

$$\int_X f(x,y)\, d\lambda^* \geqq v_2, \qquad y \in Y. \tag{29}$$

Since $f(x,y)$ is concave on the compact set X for a fixed $y \in Y$, and since λ^* is a nonnegative measure with $\lambda^*(X) = 1$, we have, by inequality (21),

(30) $$f(x_{\lambda^*}, y) \geqq \int_X f(x,y)\,d\lambda^*, \quad x_{\lambda^*} = \int_X x\,d\lambda^* \in X.$$

We obtain from relations (29) and (30)

$$v_2 \leqq \int_X f(x,y)\,d\lambda^* \leqq f(x_{\lambda^*}, y)$$

for any $y \in Y$ and some $x_{\lambda^*} \in X$, i.e.,

(31) $$v_2 \leqq \inf_{y \in Y} f(x_{\lambda^*}, y) \leqq \sup_X \inf_Y f(x,y) = v_1.$$

But by inequality (20) (Lemma 1) we have $v_1 \leqq v_2$. Therefore,

$$v_1 = \sup_X \inf_Y f(x,y) = v_2 = \inf_Y \sup_X f(x,y).$$

Thus Theorem 1 has been proved for $v_2 > -\infty$. If $v_2 = -\infty$, then by Lemma 1 we have $v_1 \leqq v_2 = -\infty$, which proves the theorem also in this case. The theorem has been completely proved.

Our further efforts will be directed toward weakening the requirements of Theorem 1 which guarantee that relations (19) hold. We introduce the functions

$$\varphi(x) = \inf_Y f(x,y), \quad x \in X; \qquad \psi(y) = \sup_X f(x,y), \quad y \in Y.$$

In the notation adopted earlier,

$$v_1 = \sup_X \varphi(x), \qquad v_2 = \inf_Y \psi(y).$$

We point out some of the simplest properties of the functions $\varphi(x)$ and $\psi(y)$.

LEMMA 3. *If $f(x,y)$ is concave in $x \in X$ (X is convex), then $\varphi(x)$ is concave.*

If $f(x,y)$ is convex in $y \in Y$ (Y is convex), then $\psi(y)$ is convex.

PROOF. We shall prove only the first part of the lemma (the proof of the second part is similar). Let $x_\alpha = \alpha x_1 + (1-\alpha)x_2$, where x_1 and x_2 are arbitrary points of X, and α is any number in $(1,0)$. We fix a $y \in Y$. Then, by the assumption of the lemma,

$$f(x_\alpha, y) \geqq \alpha f(x_1, y) + (1-\alpha) f(x_2, y).$$

Using this inequality, we have

$$\begin{aligned}\varphi(x_\alpha) = \inf_Y f(x_\alpha, y) &\geqq \inf_Y [\alpha f(x_1, y) + (1-\alpha) f(x_2, y)] \\ &\geqq \alpha \inf_Y f(x_1, y) + (1-\alpha) \inf_Y f(x_2, y) \\ &= \alpha\varphi(x_1) + (1-\alpha)\varphi(x_2).\end{aligned}$$

Therefore, $\varphi(x)$ is concave. The lemma has been proved.

Before we formulate the next assertion, we recall the notion of semicontinuity. One says that a function $z(t)$ is *upper* (*lower*) *semicontinuous* at a point t_0 if, for any $\epsilon > 0$, there exists a $\delta(\epsilon)$ such that, for $|t - t_0| \leqq \delta(\epsilon)$,

$$z(t) \leqq z(t_0) + \epsilon \quad (z(t) \geqq z(t_0) + \epsilon).$$

LEMMA 4. *If $f(x,y)$ is upper semicontinuous in x for any fixed $y \in Y$, then the function $\varphi(x)$ is upper semicontinuous at every point of the set X.*

If $f(x,y)$ is lower semicontinuous in y for any fixed $x \in X$, then the function $\psi(y)$ is lower semicontinuous at every point of the set Y.

PROOF. Again we restrict ourselves to the proof of that part of the lemma which concerns $\varphi(x)$.

Let $x_0 \in X$ and $\epsilon > 0$. It follows from the definition of $\varphi(x)$ that there exists a point y_ϵ such that

$$\varphi(x_0) \geqq f(x_0, y_\epsilon) - \epsilon/2.$$

Further, by virtue of the upper semicontinuity in x of $f(x, y_\epsilon)$, there exists a $\delta(\epsilon) > 0$ such that, for $|x - x_0| \leqq \delta(\epsilon)$,

$$f(x, y_\epsilon) - f(x_0, y_\epsilon) \leqq \epsilon/2.$$

Making use of both these inequalities and of the definition of $\varphi(x)$, we have

$$\varphi(x_0) \geqq f(x_0, y_\epsilon) - \epsilon/2 \geqq f(x, y_\epsilon) - \epsilon \geqq \varphi(x) - \epsilon,$$

if $|x - x_0| \leqq \delta(\epsilon)$.

The lemma has been proved.

It is well known that a function which is upper (lower) semicontinuous at every point of a compact set attains its sup (inf) on this set. Therefore, by Lemma 4, the function $\varphi(x)$ attains its supremum on any compact subset of X, and the function $\psi(y)$ attains its infimum on any compact subset of Y.

Let $G \subset X$ and $F \subset Y$. We introduce the notation

$$\varphi_F(x) = \inf_F f(x, y), \qquad \psi_G(y) = \sup_G f(x, y).$$

According to this notation,

$$\varphi(x) \equiv \varphi_Y(x) \quad \text{and} \quad \psi(y) \equiv \psi_X(y).$$

Obviously, $\varphi_{F_1}(x) \geqq \varphi_{F_2}(x)$ and $\psi_{G_1}(y) \leqq \psi_{G_2}(y)$ if $F_1 \subset F_2$ and $G_1 \subset G_2$. Let us agree to say that a function $z'(t)$ is a *majorant* (*minorant*) function of a function $z(t)$ on a subset S of its domain, if

$$z(t) < z'(t), \quad t \in S \quad (z(t) > z'(t), \ t \in S).$$

Lemma 5. *Let $f(x, y)$ be upper semicontinuous in $x \in X$. If $\varphi'(x)$ is a continuous majorant of the function $\varphi(x)$ on a compact set $G \subset X$, then there exists a set F_1, consisting of a finite number of points of Y, such that $\varphi'(x)$ is also a majorant of $\varphi_{F_1}(x)$ on G. Let $f(x, y)$ be lower semicontinuous in $y \in Y$. If $\psi'(y)$ is a continuous minorant of the function $\psi(y)$ on a compact set F, then there exists a finite subset G_1 of X such that $\psi'(y)$ is also a minorant of $\psi_{G_1}(y)$ on F.*

Proof. As usual, we restrict ourselves to the proof of the first part of the assertion of the lemma.

According to the definition of the function $\varphi'(x)$, there exists, for any point $x_0 \in G$, a point $y_0 = y(x_0) \in Y$ such that $f(x_0, y_0) < \varphi'(x_0)$. By virtue of the upper semicontinuity in x of $f(x, y_0)$ and the continuity of $\varphi'(x)$, there exists a neighborhood $\delta(x_0)$ of the point x_0 on which the last inequality remains true, i.e.,

$$f(x, y_0) < \varphi'(x), \quad \text{if } x \in \delta(x_0).$$

The compact set G turns out to be covered by the neighborhoods $\delta(x)$:

$$G \subset \bigcup_{x \in G} \delta(x).$$

Therefore, by the well-known Heine-Borel Theorem, there exists a finite set of points $x_1, x_2, \cdots, x_N \in G$ such that

$$G \subset \bigcup_{\alpha=1}^{N} \delta(x_\alpha).$$

We set $F_1 = \bigcup_{\alpha=1}^{N} \{y_\alpha\}$, where $y_\alpha = y(x_\alpha)$. Let x be an arbitrary point of the compact set G. We choose an index i such that $x \in \delta(x_i)$. Then

$$f(x, y_i) < \varphi'(x), \quad \text{where } y_i = y(x_i).$$

But $\varphi_{F_1}(x) = \min_{1 \leqq \alpha \leqq N} \varphi(x, y_\alpha) \leqq f(x, y_i)$. Therefore,

$$\varphi_{F_1}(x) < \varphi'(x).$$

The lemma has been proved.

We introduce the sets X^* and Y^* consisting of the points at which the functions $\varphi(x)$ and $\psi(y)$ attain their supremum and infimum, respectively, i.e., we set

$$X^* = \left\{x : \varphi(x) = \sup_{x' \in X} \varphi(x'),\ x \in X\right\},$$

$$Y^* = \left\{y : \psi(y) = \inf_{y' \in Y} \psi(y'),\ y \in Y\right\}.$$

If the set X^* (Y^*) is not empty, and if the function $f(x, y)$ is concave (convex) on the convex set X (Y), then the set X^* (Y^*) is convex (this can easily be verified directly with the use of Lemma 3). If $f(x, y)$ is also continuous in $x \in X$ ($y \in Y$), then X^* (Y^*) is closed (this follows from the corresponding semicontinuity of $\varphi(x)$ or $\psi(y)$).

THEOREM 2. *Let $f(x, y)$ be concave in $x \in X$ and convex in $y \in Y$, let X and Y be convex sets, let one of the sets X^* and Y^* be nonempty and bounded, let the function $f(x, y)$ be continuous in the corresponding variable, and let the corresponding set X (or Y) be closed. Then*

$$\sup_X \inf_Y f(x, y) = \inf_Y \sup_X f(x, y).$$

PROOF. We set, as before,

$$v_1 = \sup_{x \in X} \varphi(x) \quad \text{and} \quad v_2 = \inf_{y \in Y} \psi(y).$$

For the sake of definiteness, let Y^* be nonempty and bounded, let Y be closed, and let $f(x,y)$ be continuous in $y \in Y$ for any fixed $x \in X$. We assume at the beginning that the number v_2 is finite, i.e., taking into account Lemma 2,

$$-\infty \leqq v_1 \leqq v_2 < \infty. \tag{32}$$

If the set Y is bounded, then the assertion of Theorem 2 would follow from Theorem 1. We shall assume in the sequel that Y is unbounded. We choose an arbitrary $\epsilon > 0$ and introduce the sets

$$Y_\epsilon = \{y : \rho(y, Y^*) \leqq \epsilon,\ y \in Y\},$$
$$\Gamma_\epsilon = \{y : \epsilon \leqq \rho(y, Y^*) \leqq 2\epsilon,\ y \in Y\},$$

where $\rho(y, Y^*)$ is the distance from the point y to the set Y^*. Since Y^* is bounded and Y is not, both of the sets thus defined are nonempty and bounded. Thus, Y_ϵ and Γ_ϵ are nonempty compact sets.

According to the definition of Γ_ϵ,

$$\psi(y) > v_2, \quad y \in \Gamma_\epsilon,$$

i.e., v_2 is a minorant of the function $\psi(y)$ on Γ_ϵ. Therefore, by Lemma 5, there exists a finite set $G_1 \subset X$ such that

$$\psi_{G_1}(y) > v_2, \quad y \in \Gamma_\epsilon.$$

Further,

$$\psi(y) > v_2 - \epsilon, \quad y \in Y_\epsilon.$$

Therefore, there exists a finite set $G_2 \subset X$ such that

$$\psi_{G_2}(y) > v_2 - \epsilon, \quad y \in Y_\epsilon.$$

We set $G = G_1 \cup G_2 = \{x_1, x_2, \cdots, x_N\}$. Then

$$\psi_G(y) \geqq \psi_{G_1}(y) > v_2, \quad y \in \Gamma_\epsilon; \tag{33}$$

$$\psi_G(y) \geqq \psi_{G_2}(y) > v_2 - \epsilon, \quad y \in Y_\epsilon. \tag{34}$$

We shall show that $\psi_G(y) > v_2$ for any $y \notin Y_\epsilon$. If $y \in \Gamma_\epsilon$, this follows from (33). Now let $\rho(y, Y^*) > 2\epsilon$. We choose a point

$y^* \in Y^*$ and consider the interval Δ which joins the points y^* and y. By virtue of the continuity of the function $\rho(y, Y^*)$, there exists a point $y' \in \Gamma_\epsilon$ which lies in Δ. If we assume that $\psi_G(y) < v_2$, then, since $\psi_G(y^*) \leqq \psi(y^*) = v_2$ and since $\psi_G(y)$ is convex on Y (see Lemma 3),

$$\psi_G(y') \leqq \alpha\psi_G(y^*) + (1-\alpha)\psi_G(y)$$

for $0 < \alpha < 1$, i.e., we obtain the relation

$$\psi_G(y') \leqq v_2, \quad y' \in \Gamma_\epsilon,$$

which is incompatible with (33) and therefore proves our assertion. Thus, $\psi_G(y) > v_2$, if $y \notin Y_\epsilon$. On the other hand, $\psi_G(y) \leqq \psi(y) = v_2$ on Y^*. Therefore, for any set $F \supset Y_\epsilon$,

$$\inf_{y \in F} \psi_G(y) = \inf_{y \in Y_\epsilon} \psi_G(y). \tag{35}$$

We choose a sequence $F_1 \subset F_2 \subset \cdots \subset F_k \subset \cdots$ of compact sets such that $Y_\epsilon \subset F_1$ and $\bigcup_k F_k = Y$. By virtue of (34) and (35),

$$\inf_{y \in F_k} \psi_G(y) > v_2 - \epsilon, \quad k = 1, 2, \cdots. \tag{36}$$

We recall that

$$\psi_G(y) = \max_{1 \leqq i \leqq N} f(x_i, y).$$

Obviously, $\psi_G(y)$ can be also represented as follows:

$$\psi_G(y) = \max_{\xi \in S} \sum_{i=1}^{N} f(x_i, y)\xi_i,$$

where

$$S = \left\{ \xi = (\xi_1, \xi_2, \cdots, \xi_N) : \sum_{i=1}^{N} \xi_i = 1, \xi_i \geqq 0,\ i = 1, 2, \cdots, N \right\}.$$

We set

$$\tilde{f}(\xi, y) = \sum_{i=1}^{N} f(x_i, y)\xi_i.$$

Then (36) can be rewritten in the form

$$\inf_{y \in F_k} \sup_{\xi \in S} \tilde{f}(\xi, y) > v_2 - \epsilon, \quad k = 1, 2, \cdots.$$

The function $\tilde{f}(\xi, y)$ is linear in ξ and convex in y; and the sets S and F_k are convex and compact. Therefore, by Theorem 1,

$$\sup_{\xi \in S} \inf_{y \in F_k} \tilde{f}(\xi, y) = \inf_{y \in F_k} \sup_{\xi \in S} \tilde{f}(\xi, y) > v_2 - \epsilon, \quad k = 1, 2, \cdots. \tag{37}$$

Let $\sup_{\xi \in S} \inf_{y \in Y} \tilde{f}(\xi, y) = v$. Obviously, for any positive integer k,

$$\sup_{\xi \in S} \inf_{y \in F_k} \tilde{f}(\xi, y) \geqq v.$$

On the other hand, by Lemma 5, for any $\delta > 0$ there exists a $k(\delta)$ such that, for $k \geqq k(\delta)$,

$$\sup_{\xi \in S} \inf_{y \in F_k} \tilde{f}(\xi, y) \leqq v + \delta.$$

Thus,

$$\lim_{k \to \infty} \sup_{\xi \in S} \inf_{y \in F_k} \tilde{f}(\xi, y) = v. \tag{38}$$

We obtain from (37) and (38)

$$\sup_{\xi \in S} \inf_{y \in Y} \tilde{f}(\xi, y) = v \geqq v_2 - \epsilon. \tag{39}$$

Since $f(x, y)$ is concave in x for any $y \in Y$,

$$\tilde{f}(\xi, y) = \sum_{i=1}^{N} f(x_i, y)\xi_i \leqq f\left(\sum_{i=1}^{N} \xi_i x_i, y\right), \quad \xi \in S.$$

Therefore,

$$\inf_{y \in Y} \tilde{f}(\xi, y) \leqq \inf_{y \in Y} f(x_\xi, y), \quad x_\xi = \sum_{i=1}^{N} \xi_i x_i \in X,$$

and consequently

$$\sup_{\xi \in S} \inf_{y \in Y} \tilde{f}(\xi, y) \leqq \sup_{x \in X} \inf_{y \in Y} f(x, y) = v_1. \tag{40}$$

Comparing (39) with (40), we obtain $v_2 \leqq v_1 + \epsilon$, or, taking into account that $\epsilon > 0$ is arbitrary, $v_2 \leqq v_1$. Recalling now (32), we finally have $v_1 = v_2$.

In order to finish the proof of the theorem, it remains to consider the cases where v_2 is not finite. If $v_2 = -\infty$, the assertion of the theorem follows from Lemma 2. If $v_2 = \infty$, then $\psi(y) \equiv \infty$, i.e., $Y = Y^*$ is compact and the assertion follows from Theorem 1.

Theorem 2 has been completely proved.

§3. Duality theorems

We return to the convex programming problem (1) – (3) and try to apply Theorem 2 to it. Assuming that the assumptions of Theorem 2 hold either for the initial problem (1) – (3), or for the dual problem (9) – (11), we arrive at two distinct duality theorems.

THEOREM 3. *Let the functions $f(x)$ and $f_i(x)$, $i = 1, 2, \cdots, m$, be continuous on the set G, which is assumed to be closed, and let the set of solutions (which are not sequences) of problem (1) – (3) be nonempty and bounded. In this case, the convex programming problem (1) – (3) and its dual problem (9) – (11) are connected by the duality relation $v = \tilde{v}$.*

PROOF. Let us consider the Lagrangian functional

$$F(x, y) = f(x) + \sum_{i=1}^{m} y_i f_i(x).$$

According to the assumption of Theorem 3, $F(x, y)$ is a concave, continuous function of $x \in G$ for any $y \geqq 0$. Further, problem (1) – (3) is equivalent, by Lemma 1, to the problem

$$\varphi(x) = \inf_{y \geqq 0} F(x, y) \to \sup, \quad x \in G,$$

i.e., the suprema v and v' of the two problems are equal, and the set $X^* = \{x : \varphi(x) = v,\ x \in G\}$ coincides with the set of solutions (not sequences) of problem (1) – (3). We now apply Theorem 2, setting

$$f(x, y) = F(x, y), \quad x \in X = G, \quad y \in Y = \{y \geqq 0\}.$$

Since $f(x, y)$, besides having the necessary properties of convexity, is continuous in x, since G is closed, and since X^* is nonempty and bounded,

$$v = \sup_{x \in G} \inf_{y \geqq 0} F(x, y) = \tilde{v} = \inf_{y \geqq 0} \sup_{x \in G} F(x, y).$$

Theorem 3 has been proved.

Let us agree to say that the constraints of problem (1) – (3) satisfy the *Slater condition* [**10**], if there exists a point $x' \in G$ such that $f_i(x') > 0$ for $i = 1, 2, \cdots, m$.

THEOREM 4. *If the constraints of the convex programming problem* (1) – (3) *satisfy the Slater condition, then the duality relation* $v = \tilde{v}$ *holds. Moreover, if* $v < \infty$, *then the infimum is attained.*

PROOF. We set again $f(x,y) = F(x,y)$, $X = G$, and $Y = \{y \geqq 0\}$. The function $f(x,y)$ is linear and, therefore, continuous in y for a fixed $x \in X$, and the set Y is closed. Therefore, in order to prove the duality relation, it is sufficient, by virtue of Lemma 1 and Theorem 2, to verify that the set $Y^* = \{\psi(y) = \tilde{v}, y \geqq 0\}$ is nonempty and bounded. Let us do that. Assume that $\tilde{v} < \infty$ (the theorem is true also without this assumption, but we shall not present here the proof for the case $\tilde{v} = \infty$). Now we choose an arbitrary point $y' \geqq 0$ for which $\psi(y') < \infty$. For any $y \geqq 0$, we have

$$\psi(y) = \sup_{x \in G}\left[f(x) + \sum_{i=1}^{m} y_i f_i(x)\right] \geqq f(x') + \sum_{i=1}^{m} y_i f_i(x')$$

$$\geqq f(x') + \min_i f_i(x') \cdot \sum_{i=1}^{m} y_i = f(x') + \alpha \sum_{i=1}^{m} y_i,$$

where $\alpha = \min_i f_i(x') > 0$ and $x' \in G$ is the point which enters the Slater condition.

Let

$$y \geqq 0, \quad \sum_{i=1}^{m} y_i > \frac{\psi(y') - f(x')}{\alpha} = A;$$

then $\psi(y) > \psi(y') \geqq \tilde{v}$. Therefore, the set Y^* (if it is nonempty) is a subset of the bounded set

$$Q = \left\{ y : y \geqq 0, \sum_{i=1}^{m} y_i \leqq A \right\}.$$

Further,

$$\tilde{v} = \inf_{y \geqq 0} \psi(y) = \inf_{y \in Q} \psi(y).$$

However, $\psi(y)$ is lower semicontinuous (Lemma 4) and,

therefore, attains its infimum on the compact set Q. Thus, Y^* is a nonempty bounded set.

The theorem has been proved.

If the constraints of a problem satisfy the Slater condition, then they obviously cannot contain the linear equations $l(x) = b$, i.e., the pairs of linear inequalities $l(x) \geqq b, \; -l(x) \geqq -b$. However, the Slater condition becomes free of this drawback after a certain modification. We assume that the constraints (2) of problem (1) – (3) have the form

$$\begin{aligned} f_i(x) &\geqq 0, \quad i = 1, 2, \cdots, m; \\ l_i(x) &= b_i, \quad i = 1, 2, \cdots, s, \end{aligned} \tag{41}$$

where $l_i(x)$, $i = 1, 2, \cdots, s$, is a system of linearly independent linear forms.

If one reduces constraints (41) to the equivalent system of inequalities and formulates the problem dual to (1), (41), (3), then, after obvious transformations, the latter takes on the form

$$\psi(y) = \sup_{x \in G} \left\{ f(x) + \sum_{i=1}^{m} y_i f_i(x) + \sum_{i=1}^{s} y_{m+i}[l_i(x) - b_i] \right\}, \quad y = (y_1, y_2, \cdots, y_{m+s}), \tag{42}$$

$$\psi(y) \to \inf, \tag{43}$$

$$y_i \geqq 0, \quad i = 1, 2, \cdots, m. \tag{44}$$

We shall say that the constraints of problem (1), (41), (3) satisfy the *modified Slater condition*, if there exists a point x' such that $f_i(x') > 0$, $i = 1, 2, \cdots, m$, $l_i(x') = b_i$, $i = 1, 2, \cdots, s$, and x' is an interior point of the set G. One can verify (in approximately the same way as in the proof of Theorem 4) that the modified Slater condition implies that the set Y^* is nonempty and bounded. Therefore, the following theorem follows from Theorem 2 (taking into account Lemma 1) as a corollary.

THEOREM 5. *If the constraints of the convex programming problem* (1), (41), (3), *satisfy the modified Slater condition, then this problem is connected with the dual problem* (42) – (44) *by* *the duality relation $v = \tilde{v}$. Moreover, if $v < \infty$, the infimum in the dual problem is attained.*

If the assumption that x' is an interior point of G (for the sake of brevity, we shall subsequently call it the *interiority condition*) were not contained among the requirements of the modified Slater condition, then Theorem 5 would include, in particular, the duality theorem for linear programming. We shall present an example which shows that, even if there are only linear constraints, the interiority condition cannot, generally speaking, be omitted. We set

$$f(x) = -\exp\{-(x_1x_2)^{1/2}\},$$
$$l_1(x) - b_1 = x_2, \quad G = \{x = (x_1, x_2) : x_1 \geqq 0, x_2 \geqq 0\}.$$

The corresponding problem has the form

$$-\exp\{-(x_1x_2)^{1/2}\} \to \sup,$$
$$x_2 = 0,$$
$$x_1 \geqq 0, \quad x_2 \geqq 0.$$

It is easy to verify that $f(x)$ is concave. Therefore, the problem just formulated is a convex programming problem. Obviously, all the requirements of the modified Slater condition but the interiority condition are satisfied (the problem does not contain any nonlinear constraints). Let us verify that, in this case, the duality relation does not hold. Indeed,

$$v = \sup_{x \in R} f(x) = -1.$$

On the other hand,

$$\lim_{x_2 \to \infty} [-\exp\{-(x_1x_2)^{1/2}\} + y_1x_2] = \infty.$$

Therefore,

$$\psi(y_1) = \sup_{x_1 \geqq 0, x_2 \geqq 0} [-\exp\{-(x_1x_2)^{1/2}\} + y_1x_2] = \infty,$$

if $y_1 > 0$.

Now let $y_1 \leqq 0$. In this case,

$$-\exp\{-(x_1x_2)^{1/2}\} + y_1x_2 \leqq 0$$

for any $x_1, x_2 \geqq 0$ and, if $x_1^{(k)} = k^2$, $x_2^{(k)} = 1/k$, then

$$\lim_{k\to\infty}[-\exp\{-(x_1^{(k)}x_2^{(k)})^{1/2}\}+y_1x_2^{(k)}]=0.$$

For this reason,

$$\psi(y_1)=\sup_{x_1\geqq 0,x_2\geqq 0}[-\exp\{-(x_1x_2)^{1/2}\}+y_1x_2]=0.$$

Thus,

$$\psi(y_1)=\begin{cases}\infty & \text{if } y_1>0,\\ 0 & \text{if } y_1\leqq 0.\end{cases}$$

Therefore,

$$\tilde{v}=\inf_{y_1}\psi(y_1)=0>v=-1.$$

We shall dwell on a particular class of convex programming problems which are usually called *piecewise linear* problems. This class includes, in particular, the general linear programming problem.

We shall agree to say that $f(x)$ is a *piecewise linear concave* function, if

$$f(x)=\min_k\left\{\sum_{j=1}^n d_j^{(k)}x_j+d^{(k)}\right\},$$

i.e., if it can be represented, to within a constant term, as the minimum of a finite number of linear functions. Problem (1) – (3) is said to be a *piecewise linear programming* problem, if f and f_i, $i=1,2,\cdots,m$, are concave piecewise linear functions and if G is a polyhedral convex set.

THEOREM 6. *If* (1) – (3) *is a piecewise linear programming problem with a nonempty set of feasible solutions, then problems* (1) – (3) *and* (4) – (6) *are connected by the duality relation. In the case where* $v=\tilde{v}<\infty$, *the supremum in the initial problem and the infimum in the dual one are attained over feasible solutions of the corresponding problems.*

PROOF. The case $v=\infty$ is obvious (see Lemma 2). Therefore, we assume that $v<\infty$. The construction presented below has much in common with the method of the proof of Theorem 1. The difference is due to the linearity in y of the Lagrangian function $F(x,y)$, which makes it possible to consider only the linear functions $\varphi(y)$, each of which is determined by $m+1$ numbers. As a result, the role of the space of continuous functions is played by an $(m+1)$-dimensional space. Thus let K be the set in the $(m+1)$-dimensional space of points $z=(z_0,z_1,\cdots,z_m)$ which is defined by the following condition: $z\in K$ if and only if there exists a point $x\in G$ such that

$$z_0 \leqq f(x), \quad z_i \leqq f_i(x), \quad i=1,2,\cdots,m. \tag{45}$$

We shall show that K is a nonempty polyhedral set. First, we consider the set $\overline{K}$ of points $w=(z,x)$ which is defined by the inequalities (45) and condition (3). Since f and f_i are piecewise linear convex functions and G is a polyhedral convex set, $\overline{K}$ is the intersection of a finite number of halfspaces, i.e., a polyhedral convex set. Therefore, so is K, which is the projection of $\overline{K}$ onto the space of points z. Now we define the ray l in the $(m+1)$-dimensional space: $l=\{z:z_0\geqq v,\ z_i=0,\ i=1,2,\cdots,m\}$. Since K is closed, the point $z^*=(v,0,\cdots,0)\in K$ (hence it follows, in particular, that the supremum v in the initial problem is attained). The remaining points of l lie outside of K (otherwise there would exist feasible solutions of problem (1) – (3) at which $f(x)$ takes on values greater than v). Thus z^* is a boundary point of K. The polyhedral set K is the intersection of the halfspaces generated by a finite number of hyperplanes. We consider those of the hyperplanes which pass through z^*. Among these hyperplanes, there certainly are the hyperplanes not parallel to the z_0-axis (otherwise K would contain points of l other than z^*). Let Π be one of them and let $(\lambda_0,\lambda)=(\lambda_0,\lambda_1,\lambda_2,\cdots,\lambda_m)$ be its direction vector. As a result of our choice, Π is a support hyperplane to K at z^*, and $\lambda_0\neq 0$. This means that

$$\lambda_0 z_0 + \sum_{i=1}^{m} \lambda_i z_i \leqq \lambda_0 v, \quad z\in K. \tag{46}$$

It follows from the form of K that (46) can hold only if

$$\lambda_0 \geqq 0 \quad \text{and} \quad \lambda_i \geqq 0, \quad i=1,2,\cdots,m.$$

Thus $\lambda_0 \geqq 0$ and $\lambda = (\lambda_1, \lambda_2, \cdots, \lambda_m) \geqq 0$. We set

$$\lambda^* = \lambda/\lambda_0; \quad (z_0, z_1, \cdots, z_m) = (f(x), f_1(x), \cdots, f_m(x)), \quad x \in G.$$

Then, by (46),

$$f(x) + \sum_{i=1}^{m} \lambda_i^* f_i(x) \leqq v$$

for any $x \in G$, i.e.,

$$\psi(\lambda^*) = \sup_{x \in G} F(x, \lambda^*) \leqq v.$$

But by Lemma 2, we have $\tilde{v} \leqq v \leqq \psi(\lambda^*)$. Therefore $v = \tilde{v} = \psi(\lambda^*)$.

The theorem has been proved.

As we already mentioned, the rejection of the interiority condition in the modified Slater condition leads, generally speaking, to the violation of the duality relation. The following assertion singles out a class of convex programming problems for which the interiority condition can be deleted.

THEOREM 7. *Let the functions $f_i(x)$ in the convex programming problem (1) – (3) be concave on an open set $S \supset G$ for $i = 1, 2, \cdots, m_1 < m$ and piecewise linear for $i = m_1 + 1, \cdots, m$. Let $f(x)$ be concave on S and let G be polyhedral. If there exists a feasible solution $x^{(0)}$ of problem (1) – (3) for which $f_i(x^{(0)}) > 0$, $i = 1, 2, \cdots, m$, then problems (1) – (3) and (4) – (6) are connected by the duality relation, and, if $v = \tilde{v} < \infty$, the infimum in the dual problem is attained.*

The proof of the theorem is based on the following auxiliary assertion.

LEMMA 6. *If $f(x)$ is a concave function defined on an open convex set S, if the convex set $\Delta \subset S$, and if $\sup_{x \in \Delta} f(x) = \alpha < \infty$, then there exist a linear function $l(x)$ and a number l_0 such that*

$$f(x) \leqq l(x) + l_0, \quad x \in S, \tag{47}$$

$$\sup_{x \in \Delta} [l(x) + l_0] = \alpha. \tag{48}$$

PROOF. The following two cases are possible:

$$\text{either } \alpha = \sup_{x \in S} f(x), \text{ or } \alpha < \sup_{x \in S} f(x).$$

In the first case, obviously, the linear function $l(x) \equiv 0$ and the number $l_0 = \alpha$ satisfy requirements (47) and (48) of the lemma.

Therefore, we consider the second case, whose proof is the main part of the proof of the lemma.

We introduce the set $\Gamma = \{x : f(x) > \alpha, \ x \in S\}$. By virtue of the assumption, $\Gamma \neq \emptyset$. Since S is an open convex set and $f(x)$ is a concave functional, Γ is an open convex set. By the definition, $\Delta \cap \Gamma = \emptyset$. Therefore, according to the separation theorem, there exists a linear function $g(x) \not\equiv 0$ such that

(49) $$\sup_{x \in \Delta} g(x) \leqq \gamma \leqq \inf_{x \in \Gamma} g(x).$$

In the space $\overline{E} = E \times L$, where L is the space of all real numbers and E is the space of points x, which contains S, we define the set

$$C = \{\bar{x} = (x, x_0) : x_0 < f(x), \ x \in S\}$$

and the linear manifold (hyperplane)

$$H = \{\bar{x} = (x, x_0) : g(x) = \gamma, \ x_0 = \alpha\}.$$

Let us verify that $C \cap H = \emptyset$. Indeed, in the opposite case we have, for some point $x' \in S$,

(50) $$f(x') > \alpha, \quad g(x') = \gamma.$$

However, since $x' \in \Gamma$ together with a sufficiently small neighborhood, we have, by (49), $\gamma \leqq g(x' + x)$, if $|x| \leqq \epsilon$, where ϵ is a positive number. Or, by virtue of (50), $g(x) \geqq 0$ if $|x| \leqq \epsilon$, which is possible only if $g(x) \equiv 0$.

Thus relations (50) cannot hold, i.e., $C \cap H = \emptyset$. Moreover, C is convex. Therefore, again according to the separation theorem, there exists a linear function $\bar{g}(\bar{x}) = g'(x) + g_0 x_0 \not\equiv 0$ such that

(51) $$g'(x) + g_0 \alpha = \sigma, \qquad \text{if } g(x) = \gamma,$$

(52) $$g'(x) + g_0 x_0 \geqq \sigma, \qquad \text{if } x_0 < f(x), \quad x \in S.$$

From (52) it follows that $g_0 \leqq 0$. The case $g_0 = 0$ is impossible. Indeed, assuming the contrary and choosing an $x' \in S$ for which $g(x') = \gamma$ (such an x' certainly exists by virtue of (49)), we obtain, according to (51) and (52),

$$g'(x - x') \geqq 0 \quad \text{if } x \in S.$$

The last relation can hold only if $g'(x) \equiv 0$ (S is an open set). Thus if $g_0 = 0$, then $\bar{g}(\bar{x}) \equiv 0$, which contradicts the choice of $\bar{g}(\bar{x})$. Therefore, $g_0 \neq 0$, i.e., $g_0 < 0$. From (52) it follows that

$$f(x) \leqq (-1/g_0)g'(x) + \sigma/g_0, \quad x \in S. \tag{53}$$

If we set $g''(x) = -g'(x)/g_0$ and $\sigma' = \sigma/g_0$, we can rewrite relations (51) and (53) in the form

$$g''(x) + \sigma' = \alpha, \qquad \text{if } g(x) = \gamma, \tag{54}$$

$$g''(x) + \sigma' \geqq f(x), \quad \text{if } x \in S. \tag{55}$$

We verify that from (54) follows

$$g''(x) \equiv \gamma_1 g(x), \tag{56}$$

where γ_1 is some number. We choose a vector $x^{(1)}$ such that $g(x^{(1)}) = 1$. Then for any vector x the representation

$$x = g(x)x^{(1)} + x^{(2)}, \quad \text{where } g(x^{(2)}) = 0,$$

holds. If $x^{(3)}$ is a vector for which $g(x^{(3)}) = \gamma$, we obtain, for $x^{(4)} = x^{(2)} + x^{(3)}$,

$$x^{(2)} = x^{(4)} - x^{(3)}, \quad g(x^{(4)}) = g(x^{(3)}) = \gamma.$$

Thus

$$x = g(x)x^{(1)} + x^{(4)} - x^{(3)}.$$

By (54), $g''(x^{(4)}) = g''(x^{(3)}) = \alpha - \sigma'$. Therefore, for any vector x,

$$g''(x) = g(x)g''(x^{(1)}),$$

which is equivalent to (56) with $\gamma_1 = g''(x^{(1)})$.

We shall show that $\gamma_1 \geqq 0$. For let us suppose, to the contrary, that $\gamma_1 < 0$. According to (55), for any $x \in \Gamma$

$$g''(x) > \alpha - \sigma',$$

or, by (56),

$$g(x) < \frac{\alpha - \sigma'}{\gamma_1} = \gamma,$$

which contradicts (49) and proves our assertion.

We obtain from (49) and (56)

$$\sup_{x \in \Delta} g''(x) = \gamma_1 \sup_{x \in \Delta} g(x) \leqq \gamma_1 \gamma = \alpha - \sigma'.$$

On the other hand, by (55),

$$\sup_{x \in \Delta} g''(x) \geqq \sup_{x \in \Delta} f(x) - \sigma' = \alpha - \sigma'.$$

Thus

$$\sup_{x \in \Delta} g''(x) = \alpha - \sigma'.$$

This equality and relation (55) mean that the linear function $l(x) = g''(x)$ and the number $l_0 = \sigma'$ satisfy conditions (47) and (48) of the lemma and therefore they are the ones we have been looking for.

The lemma has been proved.

PROOF OF THEOREM 7. Without loss of generality, we shall assume that $v < \infty$. We set

$$G_1 = \{x : f_i(x) \geqq 0,\ i = m_1 + 1, \cdots, m;\ x \in G\}$$

and write down problem (1) – (3) in the equivalent form

$$f(x) \to \sup, \tag{57}$$

$$f_i(x) \geqq 0, \quad i = 1, 2, \cdots, m_1, \tag{58}$$

$$x \in G_1. \tag{59}$$

By hypothesis, problem (57) – (59) satisfies the Slater condition. Therefore, by Theorem 4, there exists a vector $(\lambda_1^*, \lambda_2^*, \cdots, \lambda_{m_1}^*) \geqq 0$ such that

$$\sup_{x \in G_1} \left[f(x) + \sum_{i=1}^{m_1} \lambda_i^* f_i(x) \right] = v. \tag{60}$$

We set

$$f'(x) = f(x) + \sum_{i=1}^{m_1} \lambda_i^* f_i(x).$$

The function $f'(x)$ is concave on the open set $S \supset G \supset G_1$. Therefore, by Lemma 6, there exist a vector l and a number l_0 such that

$$(61)\qquad \begin{aligned} &f'(x) \leqq (l, x) + l_0, \quad x \in S; \\ &\sup_{x \in G_1} f'(x) = \sup_{x \in G_1} [(l, x) + l_0]. \end{aligned}$$

Taking into account (60), we can write the last equality in the form

$$(62)\qquad \sup_{x \in G_1} [(l, x) + l_0] = v.$$

We consider the piecewise linear programming problem

$$(l, x) \to \sup,$$

$$f_i(x) \geqq 0, \quad i = m_1 + 1, \cdots, m, \quad x \in G$$

for which the set of feasible solutions is nonempty by assumption. It follows from Theorem 6 and equality (62) that there exists a vector $(\lambda_{m_1+1}^*, \cdots, \lambda_m^*) \geqq 0$ such that

$$(63)\qquad \sup_{x \in G} \left[(l, x) + \sum_{i=m_1+1}^{m} \lambda_i^* f_i(x) \right] = v - l_0.$$

Further, taking into account (61), (63), and the form of $f'(x)$, we obtain

$$\psi(\lambda^*) = \sup_{x \in G} \left[f(x) + \sum_{i=1}^{m} \lambda_i^* f_i(x) \right] \leqq v,$$

i.e.,

$$\tilde{v} = \inf_{\lambda \geqq 0} \psi(\lambda) \leqq \psi(\lambda^*) \leqq v.$$

But, by Lemma 2, $\tilde{v} \geqq v$. Thus $v = \tilde{v} = \psi(\lambda^*)$.

The theorem has been proved.

We note that the duality theorem for linear programming problems is a particular case of Theorem 7.

§4. Dual problems, problems of finding saddle points, and optimality criteria

Let $f(x, y)$ be a function of pairs of vector arguments x and y which range over finite-dimensional convex sets X and Y, respectively. As we know, the function $f(x, y)$ generates the pair of dual problems:

(64) $$\varphi(x) \to \sup, \quad x \in X;$$

(65) $$\psi(y) \to \inf, \quad y \in Y,$$

where

$$\varphi(x) = \inf_{y \in Y} f(x, y) \quad \text{and} \quad \psi(y) = \sup_{x \in X} f(x, y).$$

The pair of dual problems (64) and (65) is closely related to the problem of finding a saddle point of the function $f(x,y)$. Let us clarify this relation.

We shall call a sequence $\{x^{(k)}, y^{(k)}\}$, $x^{(k)} \in X$, $y^{(k)} \in Y$, a *generalized saddle point* of $f(x,y)$ with $x \in X$ and $y \in Y$, if

(66) $$\overline{\lim_{k \to \infty}} \sup_{x \in X} f(x, y^{(k)}) \leqq \underline{\lim_{k \to \infty}} \inf_{y \in Y} f(x^{(k)}, y).$$

Obviously,

$$\overline{\lim_{k \to \infty}} f(x^{(k)}, y^{(k)}) \leqq \overline{\lim_{k \to \infty}} \sup_{x \in X} f(x, y^{(k)}),$$

$$\underline{\lim_{k \to \infty}} f(x^{(k)}, y^{(k)}) \geqq \underline{\lim_{k \to \infty}} \inf_{y \in Y} f(x^{(k)}, y).$$

Hence from (66) we see that, if $\{x^{(k)}, y^{(k)}\}$ is a generalized saddle point, then

$$\overline{\lim_{k \to \infty}} f(x^{(k)}, y^{(k)}) \leqq \overline{\lim_{k \to \infty}} \sup_{x \in X} f(x, y^{(k)}) \leqq \underline{\lim_{k \to \infty}} \inf_{y \in Y} f(x^{(k)}, y) \leqq \underline{\lim_{k \to \infty}} f(x^{(k)}, y^{(k)}).$$

Moreover, obviously,

$$\underline{\lim_{k \to \infty}} \sup_{x \in X} f(x, y^{(k)}) \geqq \overline{\lim_{k \to \infty}} \inf_{y \in Y} f(x^{(k)}, y).$$

Therefore, conditions (66) can be rewritten in the following equivalent form:

(67) $$\lim_{k \to \infty} \sup_{x \in X} f(x, y^{(k)}) = \lim_{k \to \infty} f(x^{(k)}, y^{(k)}) = \lim_{k \to \infty} \inf_{y \in Y} f(x^{(k)}, y).$$

We assume that the generalized saddle point $\{x^{(k)}, y^{(k)}\}$ is a stationary sequence, i.e., $\{x^{(k)}, y^{(k)}\} = \{x^{(0)}, y^{(0)}\}$. From conditions (67) for a generalized saddle point, we obtain

$$\sup_{x \in X} f(x, y^{(0)}) = f(x^{(0)}, y^{(0)}) = \inf_{y \in Y} f(x^{(0)}, y),$$

or

$$f(x,y^{(0)}) \leqq f(x^{(0)},y^{(0)}) \leqq f(x^{(0)},y), \quad x \in X, \quad y \in Y,$$

i.e., the usual definition of a saddle point. Thus the notion of a generalized saddle point is a natural generalization of the notion of a saddle point to the case of sequences. The following assertion reduces the problem of the existence of a generalized saddle point to clarifying the problem of whether the duality relation is satisfied for the corresponding pair of dual problems.

THEOREM 8. *For a generalized saddle point of $f(x,y)$ with $x \in X$ and $y \in Y$ to exist, it is necessary and sufficient that problems* (64) *and* (65) *be connected by the duality relation*

$$\sup_X \varphi(x) = \inf_Y \psi(y).$$

If this holds, then any pair $\{x^{(k)}\}$ and $\{y^{(k)}\}$ of solutions of problems (64) *and* (65), *respectively, constitutes a generalized saddle point of $f(x,y)$, $x \in X$, $y \in Y$, and conversely a generalized saddle point $\{x^{(k)}, y^{(k)}\}$ of $f(x,y)$, $x \in X$, $y \in Y$, determines the solutions $\{x^{(k)}\}$ and $\{y^{(k)}\}$ of problems* (64) *and* (65), *respectively.*

PROOF. (i) Let $f(x,y)$ have a generalized saddle point with $x \in X$ and $y \in Y$, and let the sequence $\{x^{(k)}, y^{(k)}\}$ constitute this point. By (67),

$$v_2 \leqq \lim_{k \to \infty} \psi(y^{(k)}) = \lim_{k \to \infty} \varphi(x^{(k)}) \leqq v_1,$$

where $v_1 = \sup_{x \in X} \varphi(x)$ and $v_2 = \inf_{y \in Y} \psi(y)$. Further, making use of Lemma 2, we obtain

$$\lim_{k \to \infty} \varphi(x^{(k)}) = v_1 = v_2 = \lim_{k \to \infty} \psi(y^{(k)}),$$

i.e., $\{x^{(k)}\}$ and $\{y^{(k)}\}$ are solutions of problems (64) and (65), respectively, and these problems are connected by the duality relation.

(ii) Let problems (64) and (65) be connected by the duality relation and let the sequences $\{x^{(k)}\}$ and $\{y^{(k)}\}$ be solutions of the corresponding problems. This means that

$$\lim_{k \to \infty} \varphi(x^{(k)}) = v_1 = v_2 = \lim_{k \to \infty} \psi(y^{(k)}),$$

i.e., (66) holds and therefore $\{x^{(k)}, y^{(k)}\}$ is a generalized saddle point of $f(x,y)$ with $x \in X$ and $y \in Y$.

As a corollary to Theorem 8, we obtain a criterion for a saddle point.

THEOREM 9. *In order that the function $f(x,y)$ have a saddle point with $x \in X$ and $y \in Y$, it is necessary and sufficient that problems* (64) *and* (65) *be connected by the duality relation, and that the supremum in* (64) *and infimum in* (65) *be attained. A pair $\{x^{(0)}, y^{(0)}\}$ is a saddle point of $f(x,y)$, $x \in X$, $y \in Y$, if and only if $x^{(0)}$ is a solution of* (64) *and $y^{(0)}$ is a solution of* (65).

These assertions allow one to obtain the equivalent formulation of any of the duality theorems for problems (1) – (3) and (4) – (6) (in particular, of Theorems 3 – 7) in terms of a saddle point of the Lagrangian function. For example, Theorem 4 in such terms has the following form:

If the constraints (2) *of the convex programming problem* (1) – (3) *satisfy the Slater condition, and if $\{x^{(k)}\}$ is a solution of the problem, then there exists a vector $y^* \geqq 0$ such that $\{x^{(k)}, y^*\}$ is a generalized saddle point of the Lagrangian $F(x,y)$ for $x \in G$ and $y \geqq 0$.*

In order to justify this formulation, it is sufficient to note that, by virtue of Theorem 4, problems (1) – (3) and (4) – (6) are connected by the duality relation and that the vector $y^* \geqq 0$ is a solution of problem (4) – (6), and then make use of Lemma 1 and Theorem 8. Under the assumption that a solution of problem (1) – (3) exists, Theorems 4 and 5 were obtained in terms of saddle points by Kuhn and Tucker [**5**] and Uzawa [**11**], respectively. Theorem 7 was proved, under more restrictive assumptions than stated here, by G. Š. Rubinšteĭn [**9**] (again in terms of a saddle point).

So-called *optimality criteria* – necessary and sufficient conditions for an optimizing sequence to be a solution of problem (1) – (3) – are closely connected with the duality theorems.

THEOREM 10 (THE FIRST OPTIMALITY CRITERION). *Let $X = \{x^{(k)}\}$ be an optimizing sequence of problem* (1) – (3)

with $f(X) < \infty$. *For* X *to be a solution of* (1) – (3), *it is sufficient and, if the duality relation* ($v = \tilde{v}$) *holds, also necessary that there exists a sequence* $\{y^{(k)}\}$, $y^{(k)} \geqq 0$, *such that*

$$\lim_{k\to\infty} F(x^{(k)}, y^{(k)}) = \lim_{k\to\infty} \sup_{x\in G} F(x, y^{(k)}), \tag{68}$$

$$\lim_{k\to\infty} y_i^{(k)} f_i(x^{(k)}) = 0, \quad i = 1, 2, \cdots, m. \tag{69}$$

PROOF. (i) Let $X = \{x^{(k)}\}$ be a solution of problem (1) – (3), and let problems (1) – (3) and (4) – (6) be connected by the duality relation. For $\{y^{(k)}\}$, we choose a solution of the dual problem (4) – (6). By Theorem 8, the sequence $\{x^{(k)}, y^{(k)}\}$ is a generalized saddle point of the Lagrangian $F(x,y)$ for $x \in G$ and $y \geqq 0$.

Therefore (67) holds:

$$\lim_{k\to\infty} \sup_{x\in G} F(x, y^{(k)}) = \lim_{k\to\infty} F(x^{(k)}, y^{(k)}) = \lim_{k\to\infty} \inf_{y\geqq 0} F(x^{(k)}, y). \tag{70}$$

Equality (68) follows from the first equality of relation (70). Equalities (69) are consequences of the second equality of (70), since

$$\inf_{y\geqq 0} F(x^{(k)}, y) = f(x^{(k)}).$$

(ii) Let $X = \{x^{(k)}\}$ be an optimizing sequence of problem (1) – (3) such that relations (68) and (69) hold. We have from (69) and (68)

$$f(X) = \lim_{k\to\infty} f(x^{(k)}) = \lim_{k\to\infty} F(x^{(k)}, y^{(k)}) = \lim_{k\to\infty} \sup_{x\in G} F(x, y^{(k)}).$$

Further, if we denote by R, as before, the set of feasible solutions of (1) – (3), then

$$\lim_{k\to\infty} \sup_{x\in G} F(x, y^{(k)}) \geqq \lim_{k\to\infty} \sup_{x\in R} F(x, y^{(k)}) \geqq \sup_{x\in R} f(x),$$

since, for $x \in R$, we have $f_i(x) \geqq 0$, $i = 1, 2, \cdots, m$, and $y^{(k)} \geqq 0$. Thus $f(X) \geqq \sup_{x\in R} f(x)$, i.e., X is a solution of problem (1) – (3).

The theorem has been proved.

The first optimality criterion holds for any problem of the (1) – (3) type, provided that the duality relation is guaranteed to hold for this problem. In particular, the criterion holds for the classes of problems which are singled out by the assumptions of Theorems 3 – 7. If, in addition to the fact that the duality relation holds, we know that there is a solution of the dual problem which is a feasible solution (as, e.g., for problems which satisfy the assumptions of Theorems 4 – 7), then we can make the first optimality criterion more precise.

THEOREM 11 (THE SECOND OPTIMALITY CRITERION). *Let* $X = \{x^{(k)}\}$ *be an optimizing sequence of problem* (1) – (3) *with* $f(X) < \infty$. *For* X *to be a solution of this problem, it is sufficient and, if the duality relation holds and the infimum in the dual problem is attained, also necessary that there exists a vector* $y^* \geqq 0$ *such that*

$$\lim_{k\to\infty} F(x^{(k)}, y^*) = \sup_{x\in G} F(x, y^*), \tag{71}$$

$$y_i^* = 0, \quad \textit{if} \ \overline{\lim_{k\to\infty}} f(x^{(k)}) > 0. \tag{72}$$

PROOF. The necessity is proved as in Theorem 10. For y^*, one chooses a feasible solution of the dual problem which is a solution. The sufficiency follows from Theorem 10.

In particular, if $X = \{x^*\}$, then conditions (71) and (72) are transformed into

$$F(x^*, y^*) = \sup_{x\in G} F(x, y^*),$$

$$y_i^* = 0 \quad \text{if } f_i(x^*) > 0.$$

While the second criterion contains the conditions for the optimality of a feasible solution in terms of Lagrange multipliers, the first criterion states these conditions in terms of generalized Lagrange multipliers (sequences) and has a larger range of application. The optimality criteria can be rendered more concrete for particular classes of problems of the (1) – (3) type (e.g., for problems with linear constraints or with differentiable functions f and f_i). However, we shall not dwell on this.

§5. Quasiconvex problems

Until now, we have considered extremal problems with constraints whose Lagrangian functions have the property of convexity or, more generally, we have considered functions of two groups of variables which are concave in one group and convex in the second one. Later, we shall treat briefly the possibility of weakening, in some assertions, the requirements connected with the assumption that the corresponding functions are convex.

We shall begin with a definition.

A function $f(x)$, defined on a convex set X, is said to be *quasiconcave* (*quasiconvex*) if, for each real number c, the set

$$M_c = \{x : f(x) \geqq c,\ x \in X\} \quad (M_c = \{x : f(x) \leqq c,\ x \in X\})$$

is convex.

Obviously, if $f(x)$ is concave (convex), then it is also quasiconcave (quasiconvex). The opposite assertion is not true since, e.g., any monotonic function of one variable defined on an interval is both quasiconcave and quasiconvex. Therefore, the class of quasiconcave (quasiconvex) functions is broader than the class of concave (convex) functions.

Let us consider a function $f(x,y)$ of variables $x \in X$ and $y \in Y$, where, as before, X and Y are convex subsets of some Euclidean spaces. We set

$$\varphi(x) = \inf_Y f(x,y) \quad \text{and} \quad \psi(y) = \sup_X f(x,y).$$

The following theorem of H. Nikaidô [8], which is a quasiconvex analog of von Neumann's theorem (see Theorem 1), holds.

THEOREM 12. *If X and Y are compact convex sets, and if $f(x,y)$ is quasiconcave in x for any fixed $y \in Y$, quasiconvex in y for any fixed $x \in X$ and continuous in both variables x and y, then $f(x,y)$ has at least one saddle point with respect to X and Y, and*

$$\begin{aligned} \varphi(x_0) &= \max_{x \in X} \min_{y \in Y} f(x,y) \\ &= \psi(y_0) = \min_{y \in Y} \max_{x \in X} f(x,y) = f(x_0, y_0) \end{aligned} \tag{73}$$

where (x_0, y_0) is an arbitrary saddle point of $f(x,y)$ for $x \in X$ and $y \in Y$.

The proof of this theorem is based on a well-known fixed-point theorem of S. Kakutani. Before presenting the formulation of the latter, we introduce a needed notion.

Let $\Phi(z)$ be a set-valued mapping which assigns to every point z of a set Z the set $\Phi(z) \subset Z$. One says that the mapping $\Phi(z)$ is *upper semicontinuous* at a point $z_0 \in Z$ if, for any $\epsilon > 0$, there exists a $\delta > 0$ such that, for $|z - z_0| < \delta$, the set $\Phi(z)$ is contained in the ϵ-neighborhood of the set $\Phi(z_0)$.

THEOREM (S. KAKUTANI). *Let $\Phi(z)$ be a set-valued mapping of points of a convex compact set Z into its closed convex subsets. If the mapping $\Phi(z)$ is upper semicontinuous at every point $z \in Z$, then it has a fixed point, i.e., there exists a point $z_0 \in Z$ such that $z_0 \in \Phi(z_0)$.*

PROOF OF THEOREM 12. Let $Z = X \times Y$. We set, for $z = (x, y) \in Z$,

$$\Phi(z) = \left\{ x' : f(x', y) = \max_{x'' \in X} f(x'', y), x' \in X \right\}$$
$$\times \left\{ y' : f(x, y') = \min_{y'' \in Y} f(x, y''), y' \in Y \right\} = \Phi_x(z) \times \Phi_y(z).$$

By virtue of the quasiconcavity (quasiconvexity) of $f(x, y)$ in the corresponding arguments and of the continuity of $f(x, y)$, the set $\Phi(z) \subset Z$ is nonempty, closed and convex for any $z \in Z$. We shall show that the mapping $\Phi(z)$ is upper semicontinuous at every point $z \in Z$. Indeed, if this is not true, then there exist points $z^{(k)} \to z$ as $k \to \infty$ such that, for any k, the set $\Phi(z^{(k)})$ contains a point $w^{(k)}$ which lies outside of the ϵ-neighborhood of the set $\Phi(z)$ for some $\epsilon > 0$, i.e.,

$$\rho(w^{(k)}, \Phi(z)) \geqq \epsilon, \qquad w^{(k)} \in \Phi(z^{(k)}), \quad k = 1, 2, \cdots.$$

Without loss of generality, we can assume that $\{w^{(k)}\}$ is a convergent sequence (otherwise, making use of the compactness of the set Z which contains the points $w^{(k)}$, we can choose a convergent subsequence of $\{w^{(k)}\}$ and replace $\{w^{(k)}\}$ by it). Let

$$\lim_{k \to \infty} w^{(k)} = w^* = (x^*, y^*).$$

Then on the one hand

$$\rho(w^*, \Phi(z)) \geqq \epsilon > 0. \tag{74}$$

On the other hand, since $w^{(k)} = (\bar{x}^{(k)}, \bar{y}^{(k)}) \in \Phi(z^{(k)})$, where $z^{(k)} = (x^{(k)}, y^{(k)})$,

$$f(\bar{x}^{(k)}, y^{(k)}) \geqq f(x', y^{(k)}), \quad x' \in X,$$
$$f(x^{(k)}, \bar{y}^{(k)}) \leqq f(x^{(k)}, y'), \quad y' \in Y.$$

Letting k tend to infinity in these relations, we obtain

$$f(x^*, y) \geqq f(x', y), \quad x' \in X,$$
$$f(x, y^*) \leqq f(x, y'), \quad y' \in Y,$$

i.e., $w^* = (x^*, y^*) \in \Phi(z)$, $z = (x, y)$, which contradicts (74).

Therefore, the mapping $\Phi(z)$ is upper semicontinuous, and we can apply Kakutani's theorem to it. According to this theorem, there exists a point $z^* = (x^*, y^*) \in Z$ such that $z^* \in \Phi(z^*)$, i.e.,

$$f(x^*, y^*) = \max_{x \in X} f(x, y^*) = \min_{y \in Y} f(x^*, y).$$

Thus, (x^*, y^*) is a saddle point of the function $f(x, y)$ for $x \in X$ and $y \in Y$.

Now let (x_0, y_0) be an arbitrary saddle point of $f(x, y)$ for $x \in X$ and $y \in Y$, i.e.,

$$\psi(y_0) = \max_{x \in X} f(x, y_0) = f(x_0, y_0) = \min_{y \in Y} f(x_0, y) = \varphi(x_0). \tag{75}$$

But, according to Lemma 2, $\psi(y) \geqq \varphi(x)$ for any $x \in X$ and $y \in Y$. From this inequality and (75), we obtain (73).

Theorem 12 has been proved.

The basic assertion of the theorem of H. Nikaidô, like that of von Neumann's theorem, holds under assumptions on the sets X and Y which are weaker than those of Theorem 12. Since the proof of the following assertion differs only slightly from the proof of Theorem 2, we restrict ourselves only to the formulation.[1] We introduce, as before, the sets X^* and

---

[1] One can find a proof of Theorem 13 in [3].

Y^* which consist of all points at which $\varphi(x)$ attains its maximum on X and $\psi(y)$ attains its minimum on Y, respectively.

THEOREM 13. *Let a function $f(x,y)$ be continuous on $X \times Y$, quasiconcave in $x \in X$ for any fixed $y \in Y$, and quasiconvex in $y \in Y$ for any fixed $x \in X$, and let X and Y be closed convex sets. If one of the sets X^* and Y^* is nonempty and bounded, then*

$$\sup_{x \in X} \inf_{y \in Y} f(x,y) = \inf_{y \in Y} \sup_{x \in X} f(x,y).$$

In Theorem 13, one requirement is weakened in comparison with Theorem 2—instead of convexity, we assume quasiconvexity. In return, we strengthened the second requirement: in Theorem 13, the function $f(x,y)$ is assumed to be continuous in both sets of variables, while in Theorem 2 this assumption is made only with respect to the group of variables to which the nonempty bounded set X^* (or Y^*) corresponds. Theorem 13, like Theorem 2, can be used for proving duality theorems for extremal problems with constraints. However, the application of this theorem involves certain difficulties. It is worthwhile to dwell on them. We consider problem (1)–(3) with the convex set G and quasiconcave functions $f(x)$ and $f_i(x)$. Subsequently, we shall call this a *quasiconvex programming problem*. As in a convex programming problem, we can introduce the Lagrangian function

$$F(x,y) = f(x) + \sum_{i=1}^{m} y_i f_i(x), \quad x \in G, \ y \geqq 0. \tag{76}$$

By virtue of Lemma 1, the initial problem is equivalent to the following one:

$$\varphi(x) = \inf_{y \geqq 0} F(x,y) \to \sup, \quad x \in G. \tag{77}$$

If, as before, we introduce the dual problem

$$\psi(y) = \sup_{x \in G} F(x,y) \to \inf, \quad y \geqq 0,$$

then the duality theorems are reduced to establishing various conditions under which the equality

$$\sup_{x \in X} \inf_{y \in Y} F(x, y) = \inf_{y \in Y} \sup_{x \in X} F(x, y) \tag{78}$$

holds.

It would seem that, in order to prove (78), one should employ Theorem 13, However, this is impossible since $F(x, y)$ is not, generally speaking, quasiconcave in x. This is so because the class of quasiconvex functions, which includes the class of convex functions, does not have one useful property of the latter. Namely, the sum of two quasiconcave (quasiconvex) functions is not necessarily a function of the same class. One can easily convince oneself that this is indeed so from examples of functions of one variable.

Thus the Lagrangian (76) turns out to be of little use in the case of a quasiconvex programming problem. It seems that one must construct other Lagrangian functions. One should require of a Lagrangian $F(x, y)$ only the following properties:

(i) Problem (77), constructed from the Lagrangian function, must be equivalent to the initial problem (in the general case, the set $y \geqq 0$ can be replaced by some convex set Y);

(ii) $F(x, y)$ must be quasiconcave in x and quasiconvex in y.

If such a function has been constructed for some subclass of quasiconvex programming problems, then in order to obtain the duality theorems which connect the initial and dual problems it is sufficient to use Theorem 13.

As an example, we consider a fractional quasiconvex programming problem of the form

$$f(x) = \min_{1 \leqq i \leqq t} \frac{f_i'(x)}{f_i''(x)} \rightarrow \sup, \tag{79}$$

$$f_i(x) \geqq 0, \quad i = 1, 2, \cdots, m, \tag{80}$$

$$x \in G. \tag{81}$$

Here, $f_i'(x)$, $i = 1, 2, \cdots, t$, and $f_i(x)$, $i = 1, 2, \cdots, m$, are concave functions defined on a convex set G, and $f_i''(x)$, $i = 1, 2, \cdots, t$, are linear (to within a constant term) functions defined on G. We assume that

$$\min_{1 \leqq i \leqq t} f_i''(x) \geqq \alpha > 0$$

whenever $x \in G$, so that the function $f(x)$ is defined at all points of G. This function is not, generally speaking, concave. However, it is quasiconcave, since, for any c,

$$M_c = \{f(x) \geqq c, x \in G\} = \bigcap_{i=1}^{t} \{f_i'(x) - c f_i''(x) \geqq 0\} \cap G$$

is a convex set.

We set

$$F(x,y) = \frac{\sum_{i=1}^{t} y_i f_i'(x) + \sum_{i=1}^{m} y_{t+i} f_i(x)}{\sum_{i=1}^{t} y_i f_i''(x)},$$

$$y = (y_1, y_2, \cdots, y_{t+m}) \geqq 0, \quad x \in G.$$

One can verify that $F(x,y)$ satisfies conditions (i) and (ii) if, instead of all $y \geqq 0$, one considers the set

$$Y = \left\{ y : y \geqq 0, \sum_{i=1}^{t} y_i = 1 \right\}.$$

The function $F(x,y)$ is a Lagrangian for problem (79) – (81). The dual problem is to be constructed with the aid of this function.

Thus a Lagrangian function, which has form (76) in the case of a convex programming problem, should be constructed in some other way, generally speaking, if one considers quasiconvex programming problems.

§6. Generalized duality theorems

We consider a function $f(x,y)$ which is concave in $x \in X$ and convex in $y \in Y$. We introduce again the functions

$$\varphi(x) = \inf_{y \in Y} f(x,y) \quad \text{and} \quad \psi(y) = \sup_{x \in X} f(x,y).$$

As we know (Theorem 1), under the assumption that one of the convex sets X and Y is bounded and that $f(x,y)$ is continuous in the corresponding variable, we have

$$v_1 = \sup_{x \in X} \varphi(x) = v_2 = \inf_{y \in Y} \psi(y),$$

i.e., the extremal problems

$$\varphi(x) \to \sup, \quad x \in X \tag{82}$$

and

$$\psi(y) \to \inf, \quad y \in Y \tag{83}$$

are connected by the duality relation. If one rejects the assumption that one of the sets X and Y is bounded, then it can happen that $v_1 < v_2$.

We shall now construct a sequence of problems of type (82) which is, "in the limit", connected with problem (83) by the duality relation, regardless of whether the set X or Y is bounded or not.

Thus let X and Y be convex sets, let Y be closed, and let $f(x,y)$ be concave in $x \in X$, and convex and continuous in $y \in Y$. We set

$$C_k = \{y : |y| \leqq k\}, \quad Y_k = Y \cap C_k, \quad k = 1, 2, \cdots.$$

Without loss of generality, we assume that Y_k is not empty for any k.

Let

$$\varphi_k(x) = \inf_{y \in Y_k} f(x,y), \quad v_1(k) = \sup_{x \in X} \varphi_k(x), \quad k = 1, 2, \cdots.$$

We consider the family of the problems

$$\varphi_k(x) \to \sup, \quad x \in X, \quad k = 1, 2, \cdots. \tag{84}$$

We apply Theorem 1 to the function $f(x,y)$, which is defined on the set $X \times Y_k$, $k = 1, 2, \cdots$. According to this theorem,

$$v_1(k) = v_2(k) = \inf_{y \in Y_k} \psi(y). \tag{85}$$

Further, noting that $Y = \bigcup_{k=1}^{\infty} Y_k$, $Y_k \subset Y_{k+1}$, $k = 1, 2, \cdots$, we obtain

$$\lim_{k \to \infty} v_2(k) = v_2,$$

and, taking into account (85), we arrive at the following limit duality relation:

(86) $$\lim_{k\to\infty} v_1(k) = v_2.$$

Now we assume that

$$f(x,y) = F(x,y) = f(x) + \sum_{i=1}^{m} y_i f_i(x)$$

is the Lagrangian function for the convex programming problem (1) – (3) with $X = G$ and $Y = \{y : y_i \geqq 0,\ i = 1, 2, \cdots, m\}$. Since $F(x,y)$ is linear and therefore continuous in y, we have, according to (86),

(87) $$\lim_{k\to\infty} v(k) = \tilde{v},$$

where

$$v(k) = \sup_{x\in G} \varphi_k(x), \qquad \varphi_k(x) = \inf_{y\in Y_k} F(x,y),$$

$$\tilde{v} = \inf_{y\geqq 0} \psi(y).$$

Making use of the particular forms of the Lagrangian $F(x,y)$ and the set Y_k, one can perform explicitly the "inf" operation which occurs in the definition of the function $\varphi_k(x)$. For any real number a, we set

$$a^- = \begin{cases} -a, & \text{if } a \leqq 0, \\ 0, & \text{if } a > 0. \end{cases}$$

In this event,

(88) $$\varphi_k(x) = f(x) - k\sqrt{\sum_{i=1}^{m} [f_i^-(x)]^2}, \quad x \in G.$$

Indeed, let $I(x) = \{i : f_i(x) < 0\}$. If $I(x) = \emptyset$, then, on the one hand,

$$\varphi_k(x) \leqq F(x,0) = f(x)$$

and, on the other hand, $F(x,y) \geqq f(x)$ for any $y \geqq 0$, i.e., $\varphi_k(x) \geqq f(x)$.

Thus, for $I(x) = \emptyset$,

$$\varphi_k(x) = f(x),$$

which corresponds to formula (88).

We assume now that $I(x) \neq \emptyset$. For any $y \in Y_k$ $(y \geqq 0, |y| \leqq k)$, we have

$$F(x,y) = f(x) + \sum_{i=1}^{m} y_i f_i(x) \geqq f(x) - \sum_{i=1}^{m} y_i f_i^-(x).$$

Further, using the Cauchy-Bunjakovskiĭ inequality, we obtain

$$F(x,y) \geqq f(x) - \sqrt{\sum_{i=1}^{m} y_i^2} \cdot \sqrt{\sum_{i=1}^{m} [f_i^-(x)]^2}$$
$$\geqq f(x) - k \sqrt{\sum_{i=1}^{m} [f_i^-(x)]^2}.$$

Therefore

$$\varphi_k(x) = \inf_{y \in Y_k} F(x,y) \geqq f(x) - k \sqrt{\sum_{i=1}^{m} [f_i^-(x)]^2}. \tag{89}$$

On the other hand, if we set

$$y_i(x) = \frac{k f_i^-(x)}{\sqrt{\sum_{i=1}^{m} [f_i^-(x)]^2}}, \quad i = 1, 2, \cdots, m,$$

then the vector $y(x) = (y_1(x), y_2(x), \cdots, y_m(x))$ belongs to Y_k, which means that

$$\varphi_k(x) \leqq F(x, y(x)) = f(x) - k \sqrt{\sum_{i=1}^{m} [f_i^-(x)]^2}. \tag{90}$$

Equality (88) is a consequence of (89) and (90).

We introduce the notion of a generalized initial problem. We shall say that a sequence $X = \{x^{(k)}\}$, $x^{(k)} \in G$, is a *generalized feasible solution* of problem (1) – (3), if

(i) the following limit exists, finite or infinite:

$$\lim_{k \to \infty} f(x^{(k)}) = f(X),$$

(ii) $\underline{\lim}_{k \to \infty} f_i(x^{(k)}) \geqq 0$ for any $i = 1, 2, \cdots, m$.

We denote the set of all generalized feasible solutions of problem (1) – (3) by $\overline{R}$, and we set

$$\bar{v} = \begin{cases} \sup_{X \in \overline{R}} f(X), & \text{if } \overline{R} \neq \emptyset, \\ -\infty, & \text{if } \overline{R} = \emptyset. \end{cases}$$

We shall call the problem of maximizing the function $f(X)$ on the set of generalized feasible solutions the *generalized* problem (1) – (3). The following generalized duality theorem is true.

THEOREM 14. *Let* (1) – (3) *be a convex programming problem with* $\tilde{v} < \infty$. *Then the generalized problem* (1) – (3) *and dual problem* (9) – (11) *are connected by the duality relation, i.e.*,

$$\bar{v} = \tilde{v}. \tag{91}$$

PROOF. (i) Let $X = \{x^{(k)}\}$ be a generalized feasible solution of problem (1) – (3). Using (88) and properties (i) and (ii) of the sequence X we have for any t

$$\lim_{k\to\infty} \varphi_t(x^{(k)}) = \lim_{k\to\infty}\left[f(x^{(k)}) - t\sqrt{\sum_{i=1}^{m}[f_i^-(x^{(k)})]^2}\right] = f(X).$$

Therefore

$$v(t) = \sup_{x\in G}\varphi_t(x) \geqq \sup_{x\in\bar{R}} f(X) = \bar{v}, \quad t = 1, 2, \cdots.$$

Further we obtain, taking into account equality (87),

$$\tilde{v} = \lim v(t) \geqq \bar{v}. \tag{92}$$

If $\bar{R} = \emptyset$, then $\bar{v} = -\infty$, and inequality (92) also holds. If $\tilde{v} = -\infty$, then by (92), $\bar{v} = \tilde{v} = -\infty$. Therefore, we shall henceforth assume that $\tilde{v} > -\infty$.

(ii) It follows from (87) that there exists a sequence $\bar{X} = \{\bar{x}^{(k)}\}$, $\bar{x}^{(k)} \in G$, such that

$$\lim_{k\to\infty}\varphi_k(\bar{x}^{(k)}) = v.$$

Let

$$I(x) = \{i : f_i(x) < 0,\ i = 1, 2, \cdots, m\};$$
$$y(x) = (y_1(x), y_2(x), \cdots, y_m(x)),$$

where

$$y_i(x) = \begin{cases} 0, & \text{if } I(x) = \emptyset, \\ \dfrac{f_i^-(x)}{\sqrt{\sum_{i=1}^{m}[f_i^-(x)]^2}}, & \text{if } I(x) \neq \emptyset. \end{cases}$$

We choose a vector $y_0 \geqq 0$ such that $\psi(y_0) < \infty$ (the existence of such a vector follows from the assumption that $v < \infty$). Let $t > |y_0|$. Then

$$\bar{y} = y_0 + (t - |y_0|)\, y(x) \geqq 0,$$

$$|\bar{y}| \leqq |y_0| + (t - |y_0|)\, |y(x)| \leqq t,$$

i.e., $\bar{y} \in Y_t$. Therefore

$$\varphi_t(x) = \inf_{y \in Y_t} F(x,y) \leqq F(x,\bar{y})$$

$$= F(x,y_0) - (t - |y_0|) \sqrt{\sum_{i=1}^{m} [f_i^-(x)]^2}.$$

Obviously, $\psi(y_0) \geqq F(x,y_0)$, and the inequality thus obtained leads to the relation

$$\sqrt{\sum_{i=1}^{m} [f_i^-(x)]^2} \leqq \frac{\psi(y_0) - \varphi_t(x)}{t - |y_0|}, \quad t > |y_0|. \tag{93}$$

We substitute $\bar{x}^{(y)}$ for x in (93) and let t tend to infinity, obtaining

$$\lim_{t\to\infty} \sqrt{\sum_{i=1}^{m} [f_i^-(\bar{x}^{(t)})]^2} \leqq (\psi(y_0) - \tilde{v}) \cdot \lim_{t\to\infty} \frac{1}{t - |y_0|} = 0,$$

i.e.,

$$\varliminf_{t\to\infty} f_i(\bar{x}^{(t)}) \geqq \lim_{t\to\infty} [-f_i^-(\bar{x}^{(t)})] = 0 \tag{94}$$

for $i = 1, 2, \cdots, m$. Since $\varphi_k(\bar{x}^{(t)}) \leqq F(\bar{x}^{(t)}, 0) = f(\bar{x}^{(t)})$, we have

$$\varliminf_{t\to\infty} f(\bar{x}^{(t)}) \geqq \lim_{t\to\infty} \varphi_k(\bar{x}^{(t)}) = \tilde{v}. \tag{95}$$

Further, taking into account (94), we obtain

$$\varliminf f(\bar{x}^{(t)}) \leqq \varlimsup f(\bar{x}^{(t)}) \leqq \bar{v}. \tag{96}$$

From (95) and (96), it follows that

$$\bar{v} \geqq \tilde{v}. \tag{97}$$

Inequalities (97) and (92) lead to the desired equality (91).

We note that from (91), (96), and (97) follows the existence of

$$\lim_{t\to\infty} f(\bar{x}^{(t)}) = f(\bar{X})$$

with $f(\bar{x}) = \bar{v}$, i.e., $\overline{X}$ is an optimal generalized feasible solution of problem (1) – (3).

The generalized duality theorem was proved here under the assumption that $\tilde{v} < \infty$. If $\tilde{v} = \infty$, one can show that, for $\overline{R} \neq \emptyset$, equality (91) also holds. Thus if the generalized duality relation turns out to be violated for a convex programming problem, then $\overline{R}$ does not contain any elements and $\tilde{v} = \infty$. This possibility of violating the generalized duality relation can be realized even in the case of linear programming (it means that the initial and dual linear programming problems have empty sets of feasible solutions).

§7. Stability and marginal values

In conclusion, we shall consider problems which are connected with the continuous dependence of the sets of optimal solutions to the initial and dual problems on a parameter. We shall obtain conditions which guarantee that the family of these sets is continuous (stability theorem). As an application of the stability theorem, we shall derive formulas for the marginal values of a convex programming problem.

Let $f(x,y,t)$ be a function defined and continuous on a set $X \times Y \times T$, quasiconcave in x for all values of y, t in $Y \times T$, and quasiconvex in y for all values of x, t in $X \times T$. Here X and Y are closed convex subsets of Euclidean spaces, and T is a metric space (in particular, T may be a Euclidean space).

We set

$$\varphi(x,t) = \inf_{y \in Y} f(x,y,t),$$

$$\psi(y,t) = \sup_{x \in X} f(x,y,t).$$

We consider the family of pairs of dual problems

$$\varphi(x,t) \to \sup, \quad x \in X, \tag{98}$$

$$\psi(y,t) \to \inf, \quad y \in Y. \tag{99}$$

Let

$$v_1(t) = \sup_{x \in X} \varphi(x,t) \quad \text{and} \quad v_2(t) = \inf_{y \in Y} \psi(y,t).$$

We introduce the sets of all solutions to problems (98) and (99):

$$X(t) = \{x : \varphi(x,t) = v_1(t),\ x \in X\},$$
$$Y(t) = \{y : \psi(y,t) = v_2(t),\ y \in Y\}.$$

The following theorem on the stability of the sets $X(t)$ and $Y(t)$ at a given point is valid.

THEOREM 15. *If for some limit point $q \in T$ the sets $X(q)$ and $Y(q)$ are nonempty and bounded, then there exists a neighborhood $\delta(q) \subset T$ of the point q such that, for all $t \in \delta(q)$, the sets $X(t)$ and $Y(t)$ are nonempty, $v_1(t) = v_2(t)$, and the set-valued mappings of $t \in \delta(q)$ into $X(t)$ and $Y(t)$ are upper semicontinuous at q.*

PROOF. (i) First of all, we note that Theorem 13 implies

$$\inf_{y \in Y} \sup_{x \in X} f(x,y,q) = \sup_{x \in X} \inf_{y \in Y} f(x,y,q) = f(x_q, y_q, q),$$

where (x_q, y_q) is an arbitrary point of $X(q) \times Y(q)$.

We set

$$v(q) = \max_{x \in X} \varphi(x,q) = \min_{y \in Y} \psi(y,q). \tag{100}$$

We choose an arbitrary $\sigma > 0$ and denote by

$$X_\sigma = \{x : x \in X,\ \rho(x, X(q)) \leqq \sigma\}$$

the set of all points which belong to X and are no farther from $X(q)$ than σ.

We shall show that, for σ sufficiently large,

$$\varliminf_{t \to q} \sup_{x \in X_\sigma} \varphi(x,t) \geqq v(q). \tag{101}$$

We shall argue by contradiction. We assume that the statement is not true, and we fix an arbitrary $\sigma > 0$. Then there exist a $\delta > 0$ and a sequence $t_k \to q$ such that

$$\sup_{x \in X_\sigma} \varphi(x, t_k) < v(q) - \delta \text{ for all } k.$$

From Theorem 13 it follows that

$$\sup_{x \in X_\sigma} \varphi(x, t_k) = \inf_{y \in Y} \sup_{x \in X_\sigma} f(x,y,t_k) < v(q) - \delta,$$

i.e., there exists a sequence of y_k such that

(102) $$\sup_{x \in X_\sigma} f(x, y_k, t_k) < v(q) - \delta \text{ for all } k.$$

We shall show that $|y_k| \to \infty$ as $k \to \infty$. Indeed, in the opposite case we can assume without loss of generality that $y_k \to y^*$, $y^* \in Y$, i.e.,

$$\sup_{x \in X_\sigma} f(x, y^*, q) \leqq v(q) - \delta.$$

Taking into account (100), we obtain from the last inequality the impossible relation

$$\begin{aligned} v(q) &= \sup_{x \in X_\sigma} \inf_{y \in Y} f(x, y, q) \\ &= \inf_{y \in Y} \sup_{x \in X_\sigma} f(x, y, q) \leqq \sup_{x \in X_\sigma} f(x, y^*, q) \leqq v(q) - \delta, \end{aligned}$$

which proves that our assertion is true.

Now we consider an arbitrary point $y_0 \in Y(q)$. For this point,

$$\sup_{x \in X_\sigma} f(x, y_0, q) = v(q).$$

Since the right-hand side of this equality is continuous in $t = q$, there exists for each $\epsilon > 0$ a positive integer $k(\epsilon)$ such that for all $k \geqq k(\epsilon)$

(103) $$\sup_{x \in X_\sigma} f(x, y_0, t_k) < v(q) + \epsilon.$$

For an arbitrary fixed $l > |y_0|$ we define the sequence

$$y_k^{(l)} = \frac{l}{|y_k|} y_k + \left(1 - \frac{l}{|y_k|}\right) y_0, \qquad |y_k| \geqq l,$$

which has infinitely many elements, since $|y_k| \to \infty$ as $k \to \infty$. Obviously,

(104) $$l - |y_0| \leqq |y_k^{(l)}| \leqq l + |y_0|.$$

It follows from the quasiconvexity in $y \in Y$ of the function $\sup_{x \in X_\sigma} f(x, y, t_k)$ and from relations (102) and (103) that

(105) $$\sup_{x \in X_\sigma} f(x, y_k^{(l)}, t_k) < v(q) + \epsilon.$$

Without loss of generality, we can assume that the sequence $\{y_k^{(l)}\}$ converges to the element $y^{(l)}$ (see (104)).

We emphasize that $y^{(l)}$ depends on σ, setting explicitly $y^{(l)} = y^{(l)}(\sigma)$. Then we obtain from inequality (105)

$$\sup_{x \in X_\sigma} f(x, y^{(l)}(\sigma), q) \leqq v(q) + \epsilon,$$

or since $\epsilon > 0$ is arbitrary

$$\sup_{x \in X_\sigma} f(x, y^{(l)}(\sigma), q) \leqq v(q). \tag{106}$$

We obtain from (104)

$$l - |y_0| \leqq |y^{(l)}(\sigma)| \leqq l + |y_0|. \tag{107}$$

According to (107) we can assume without loss of generality that

$$\lim_{\sigma \to \infty} y^{(l)}(\sigma) = \bar{y}^{(l)}.$$

Then we have from (106)

$$\psi(\bar{y}^{(l)}, q) = \sup_{x \in X} f(x, \bar{y}^{(l)}, q) \leqq v(q). \tag{108}$$

It follows from (108) that $\bar{y}^{(l)} \in Y(q)$ for any $l \geqq |y_0|$. But, according to (107), $|\bar{y}^{(l)}| \geqq l - |y_0|$.

Thus we have obtained the statement that $Y(q)$ contains elements with arbitrary large norms, which contradicts the fact that $Y(q)$ is bounded.

Therefore, assertion (101) has been proved.

(ii) We have proved that for some $\sigma > 0$ and an arbitrary $\epsilon > 0$ there exists a neighborhood $\delta'_\epsilon(q)$ of the point q such that

$$\sup_{x \in X_\sigma} \varphi(x, t) \geqq v(q) - \epsilon, \quad t \in \delta'_\epsilon(q). \tag{109}$$

We shall now prove that there exist a neighborhood $\delta_x(q) \subset \delta'_\epsilon(q)$ of q and an $\epsilon > 0$ such that

$$\varphi(x, t) < v(q) - \epsilon \text{ for all } x \notin X_\sigma, \ t \in \delta_x(q). \tag{110}$$

We introduce the set $\Gamma_\sigma = \{x : x \in X, \sigma \leqq \rho(x, X(q)) \leqq 2\sigma\}$. By virtue of Lemma 4, the function $\varphi(x, q)$ is upper semicontinuous in x and therefore

$$\max \varphi(x, q) = \varphi(x', q) \text{ for some } x' \in \Gamma_\sigma.$$

Hence, taking into account that $X(q) \cap \Gamma_\sigma = \emptyset$, we obtain

$$\max_{x \in \Gamma_\sigma} \varphi(x,q) < v(q) - \epsilon \tag{111}$$

for some $\epsilon > 0$.

According to Lemma 5, we can choose a finite set $F \subset Y$ such that (111) implies the inequality

$$\min_{x \in \Gamma_\sigma} f(x,y,q) < v(q) - \epsilon, \quad x \in \Gamma_\sigma.$$

By continuity, this inequality remains in force for some neighborhood $\delta_x(q) \subset \delta'_\epsilon(q)$ of the point q, i.e.,

$$\varphi(x,t) \leqq \min_{y \in F} f(x,y,t) < v(q) - \epsilon, \; x \in \Gamma_\sigma, \; t \in \delta_x(q). \tag{112}$$

We shall show that, for $\delta_x(q)$ and $\epsilon > 0$ thus chosen, inequality (110) holds. Indeed, if for some $x' \notin X_\sigma$ and $t' \in \delta_x(q)$ this inequality is violated, i.e., $\varphi(x',t') > v(q) - \epsilon$, then, joining a point $x'' \in X_\sigma$ at which $\varphi(x'',t') \gneqq v(q) - \epsilon$ with the point $x' \notin X_\sigma$ by the segment S, we obtain, taking into account that $\varphi(x,t')$ is quasiconcave,

$$\varphi(x,t') \geqq v(q) - \epsilon \tag{113}$$

for any $x \in X$. Obviously S has points in common with the set Γ_σ; let $x_0 \in S \cap \Gamma_\sigma$. Then by (113)

$$\varphi(x_0,t') \geqq v(q) - \epsilon, \; x_0 \in \Gamma_\sigma, \; t' \in \delta_x(q),$$

which contradicts (112).

It follows from (109) and (110) that the set $X(t)$ coincides, for $t \in \delta_x(q)$, with the set of points at which $\sup_{x \in X_\sigma} \varphi(x,t)$ is attained. This last set is nonempty by virtue of Lemma 4 (X_σ is compact). Thus for any $t \in \delta_x(q)$, $X(t)$ is nonempty and moreover

$$X(t) \subset X_\sigma. \tag{114}$$

The similar fact for the sets $Y(t)$, $t \in \delta_y(q)$, can be proved in exactly the same way.

Thus if $t \in \delta(q) = \delta_x(q) \cap \delta_y(y)$, then in the first place the sets $X(t)$ and $Y(t)$ are nonempty and secondly $v_1(t) = v_2(t)$

in accordance with Theorem 13 (see (114)).

(iii) In order to prove that the mappings $X(t)$ and $Y(t)$ are upper semicontinuous at $t = q$, we use the inequalities

$$f(x, y_t, t) \leqq f(x_t, y_t, t) \leqq f(x_t, y, t), \tag{115}$$

where $x \in X$, $y \in Y$, $x_t \in X_t$, $y_t \in Y_t$, and $t \in \delta(q)$.

If the last assertion of the theorem is not true, e.g., with respect to the mapping $X(t)$, $t \in T$, then there exist sequences $t_k \rightarrow q$, $t_k \in \delta(q)$, and $x_k \in X(t_k)$ such that

$$\rho(x_k, X(q)) \geqq \epsilon > 0 \quad \text{for } k = 1, 2, \cdots. \tag{116}$$

Let y_k be an arbitrary point of $Y(t_k)$. Since the sets $X(t_k)$ and $Y(t_k)$ are uniformly bounded (see (114)), we can assume without loss of generality that $x_k \rightarrow x^*$ and $y_k \rightarrow y^*$ as $k \rightarrow \infty$. Then from (115) it follows that

$$f(x, y^*, q) \leqq f(x^*, y^*, q) \leqq f(x^*, y, q),$$

where $x \in X$ and $y \in Y$, i.e., $x^* \in X(q)$, which contradicts (116).

Theorem 15 has been completely proved.

Theorem 15 can be used in deriving formulas for the so-called marginal values of extremal problems.

We consider a function $f(x, y, t)$ defined on the set $X \times Y \times T$, where $T = [0, t_0]$. If we assume that the assumptions of Theorem 15 are satisfied ($q = 0$), then, according to this theorem, for some $t' > 0$ the sets $X(t)$ and $Y(t)$ are nonempty, where $0 \leqq t \leqq t'$, and

$$\varphi(t) = \max_{x \in X} \min_{y \in Y} f(x, y, t) = \min_{y \in Y} \max_{x \in X} f(x, y, t) = f(x_t, y_t, t),$$

where $x_t \in X(t)$, $y_t \in Y(t)$, and $0 \leqq t \leqq t'$.

The assertion presented below contains both the conditions which guarantee that $\varphi(t)$ is differentiable and the formula for the derivative of this function. The number $\varphi'(0)$ is usually called the *marginal value* of the family of problems (98) and (99) for $t = 0$, and the corresponding assertion is called the *theorem on marginal values*.

THEOREM 16. *Let the assumptions of Theorem* 15 *be satisfied, let the function* $f(x,y,t)$ *be right differentiable with respect to* t *at the point* $t=0$ *for any* (x,y) *in some neighborhood* Ω *of the set* $X(0)\times Y(0)$, *and let the derivative* $f_t'(x,y,0)$ *of this function satisfy the condition*

$$\frac{f(x,y,t)-f(x,y,0)}{t}\to f_t'(x,y,0) \quad as \quad t\to +0$$

uniformly with respect to $(x,y)\in\Omega$.

Then the right derivative $\varphi'(0)$ *of the function* $\varphi(t)$ *at the point* $t=0$ *exists and moreover*

$$\varphi'(0)=\max_{x\in X(0)}\min_{y\in Y(0)} f_t'(x,y,0)=\min_{y\in Y(0)}\max_{x\in X(0)} f_t'(x,y,0). \tag{117}$$

PROOF. Let $0\leqq t\leqq t'$. Then the sets $X(t)$ and $Y(t)$ are nonempty and

$$f(x_0,y_t,t)\leqq f(x_t,y_t,t)\leqq f(x_t,y_0,t), \tag{118}$$

where $(x_0,y_0)\in X(0)\times Y(0)$ and $(x_t,y_t)\in X(t)\times Y(t)$.

We set

$$\mu(x,y,t)=f(x,y,t)-f(x,y,0)-tf_t'(x,y,0), \qquad (x,y)\in\Omega.$$

By the hypothesis of the theorem, we have

$$|\mu(x,y,t)|\leqq\mu(t), \quad (x,y)\in\Omega, \tag{119}$$

where $\mu(t)$ does not depend on x and y and is such that

$$\lim_{t\to+0}\frac{1}{t}\mu(t)=0. \tag{120}$$

Without loss of generality, we can assume that $X(t)\times Y(t)\subset\Omega$ whenever $t\leqq t'$ (since the corresponding set-valued mapping is upper semicontinuous). Therefore (118) can be rewritten in the form

$$\begin{aligned} f(x_0,y_t,0)+tf_t'(x_0,y_t,0)-\mu(t) &\leqq f(x_t,y_t,t)\\ &\leqq f(x_t,y_0,0)+tf_t'(x_t,y_0,0)+\mu(t). \end{aligned} \tag{121}$$

Since $(x_0,y_0)\in X(0)\times Y(0)$,

$$f(x_t, y_0, 0) \leqq f(x_0, y_0, 0) \leqq f(x_0, y_t, 0).$$

Taking into account these inequalities and (121), we obtain the following estimates of the difference quotient of $\varphi(t)$:

$$\text{(122)} \qquad f_t'(x_0, y_t, 0) - \frac{\mu(t)}{t} \leqq \frac{\varphi(t) - \varphi(0)}{t} \leqq f_t'(x_t, y_0, 0) + \frac{\mu(t)}{t}.$$

Estimates (122) are derived for arbitrary saddle points $(x_0, y_0) \in X(0) \times Y(0)$ and $(x_t, y_t) \in X(t) \times Y(t)$. Therefore from (122) follows

$$\text{(123)} \qquad \begin{aligned} \min_{y \in Y(t)} \max_{x \in X(0)} f_t'(x, y, 0) - \frac{\mu(t)}{t} &\leqq \frac{\varphi(t) - \varphi(0)}{t} \\ &\leqq \max_{x \in X(t)} \min_{y \in Y(0)} f_t'(x, y, 0) + \frac{\mu(t)}{t}. \end{aligned}$$

As before, let X_ϵ and Y_ϵ be ϵ-neighborhoods of the sets $X(0)$ and $Y(0)$, respectively. For any $t > 0$, we denote by $\epsilon(t)$ the smallest of the numbers ϵ for which $X(t) \subset X_\epsilon$ and $Y(t) \subset Y_\epsilon$. According to Theorem 15,

$$\text{(124)} \qquad \lim_{t \to +0} \epsilon(t) = 0.$$

In the notation introduced above, estimates (123) lead us to the inequalities

$$\text{(125)} \qquad \begin{aligned} \min_{y \in Y_{\epsilon(t)}} \max_{x \in X(0)} f_t'(x, y, 0) - \frac{\mu(t)}{t} &\leqq \frac{\varphi(t) - \varphi(0)}{t} \\ &\leqq \max_{x \in X_{\epsilon(t)}} \min_{y \in Y(0)} f_t'(x, y, 0) + \frac{\mu(t)}{t}. \end{aligned}$$

Relations (119) and (120) imply the continuity of $f_t'(x, y, 0)$ on Ω. Therefore the functions

$$\max_{x \in X(0)} f_t'(x, y, 0) \quad \text{and} \quad \min_{y \in Y(0)} f_t'(x, y, 0)$$

are also continuous, and thus

$$\text{(126)} \qquad \begin{aligned} \lim_{t \to +0} \min_{y \in Y_{\epsilon(t)}} \max_{x \in X(0)} f_t'(x, y, 0) &= \min_{y \in Y(0)} \max_{x \in X(0)} f_t'(x, y, 0), \\ \lim_{t \to +0} \max_{x \in X_{\epsilon(t)}} \min_{y \in Y(0)} f_t'(x, y, 0) &= \max_{x \in X(0)} \min_{y \in Y(0)} f_t'(x, y, 0). \end{aligned}$$

Estimates (125) and relations (120), (124), and (126) lead to the inequalities

$$\varlimsup_{t\to+0} \frac{\varphi(t)-\varphi(0)}{t} \leqq \max_{x\in X(0)} \min_{y\in Y(0)} f_t'(x,y,0),$$

$$\varliminf_{t\to+0} \frac{\varphi(t)-\varphi(0)}{t} \geqq \min_{y\in Y(0)} \max_{x\in X(0)} f_t'(x,y,0).$$

From these inequalities, taking into account the known inequality

$$\max_{x\in X(0)} \min_{y\in Y(0)} f_t'(x,y,0) \leqq \min_{y\in Y(0)} \max_{x\in X(0)} f_t'(x,y,0),$$

we obtain the desired inequality (117).

The theorem has been proved.

Let us consider the family consisting of the following convex programming problems, which depend on the parameter $t\in[0,t_0]$:

$$f(x,t)\to\sup, \tag{127}$$

$$f_i(x,t)\geqq 0, \quad i=1,\cdots,m, \tag{128}$$

$$x\in X, \tag{129}$$

where $f(x,t)$ and $f_i(x,t)$ are defined and continuous on a closed convex set X and concave in x, $t\in[0,t_0]$, and $i=0,1,\cdots,m$.

Let $v(t)$ be the value of the supremum in (127) under the constraints (128) and (129).

We introduce the Lagrangian functions for problems (127) – (129):

$$F(x,y,t)=f(x,t)+\sum_{i=1}^{m} y_i f_i(x,t)$$

and set $Y=\{y: y=(y_1,\cdots,y_m),\ y_i\geqq 0,\ i=1,\cdots,m\}$.

As we know, the problem dual to (127) – (129) for a fixed t has the form

$$\psi(y,t)=\sup_{x\in X} F(x,y,t)\to\inf, \tag{130}$$

$$y\in Y. \tag{131}$$

We denote problems (127) – (129) and (130), (131), for a fixed $t\in[0,t_0]$, by $A(t)$ and $B(t)$, respectively.

As a corollary to Theorem 16, we obtain the following theorem on the marginal values of problem (127) – (129).

THEOREM 17. *Let problems $A(0)$ and $B(0)$ be solvable with the bounded sets of solutions $X(0)$ and $Y(0)$, let the functions $f_0(x,t) = f(x,t)$ and $f_i(x,t)$, $i = 1, \cdots, m$, be right differentiable at the point $t = 0$ for any x in some neighborhood X' of the set $X(0)$ of solutions of the problem $A(0)$, and moreover let the derivatives $\partial f_i(x,0)/\partial t$ of these functions satisfy the condition*

$$\frac{f_i(x,t) - f_i(x,0)}{t} \longrightarrow \frac{\partial f_i(x,0)}{\partial t} \quad \text{as } t \to +0$$

uniformly with respect to $x \in X'$, $i = 0, 1, \cdots, m$.

Then the function $v(t)$ has a right derivative at the point $t = 0$, and moreover

$$v'(0) = \max_{x \in X(0)} \min_{y \in Y(0)} \frac{\partial F(x,y,0)}{\partial t} = \min_{y \in Y(0)} \max_{x \in X(0)} \frac{\partial F(x,y,0)}{\partial t}.$$

In order to prove Theorem 17, it is sufficient to note that by Lemma 1 the problem $A(t)$ can be rewritten in the following equivalent form:

$$\varphi(x,t) = \inf_{y \in Y} F(x,y,t) \to \sup, \quad x \in X,$$

and then use Theorem 16.

Theorem 17 is a generalization to convex programming problems of the well-known theorem on marginal values of linear programming problems. The latter was formulated by H. Mills, who proved it in the case of rectangular matrix games. However, the proof of this theorem offered by Mills for the case of the general linear programming problem contained a mistake. It turned out later[2)] that for the theorem on marginal values of the general linear programming problem to hold it is sufficient to assume that the sets $X(0)$ and $Y(0)$ of solutions of the initial and dual problems for $t = 0$ are bounded. It was also proved that this requirement is essential.

[2)] In this connection, see [2], Chapter III, §5.

Thus Theorem 17 has been proved under extremely weak assumptions, and it contains, as a particular case, the theorem on the marginal values of a linear programming problem.

Bibliography

[1] E. G. Gol′šteĭn, *Dual problems of convex programming,* Èkonom. i Mat. Metody **1** (1965), 410-424. (Russian) RŽMat. **1966** #1B279.

[2] E. G. Gol′šteĭn and D. B. Judin, *New directions in linear programming,* "Sovetskoe Radio", Moscow, 1966. (Russian) Zbl. **158**, 186.

[3] E. G. Gol′šteĭn and S. M. Movšovič, *The continuous dependence on a parameter of the set of solutions of a minimax problem,* Èkonom. i Mat. Metody **4** (1968), 920-930. (Russian) RŽMat. **1969** #5B447.

[4] H. Kneser, *Sur un théorème fondamental de la théorie des jeux,* C. R. Acad. Sci. Paris **234** (1952), 2418-2420. MR **14**, 301.

[5] H. W. Kuhn and A. W. Tucker, *Nonlinear programming,* Proc. Second Berkeley Sympos. Math. Statist. and Probability, Univ. of California Press, Berkeley, Calif., 1951, pp. 481-492. MR **13**, 855.

[6] H. D. Mills, "Marginal values of matrix games and linear programs", in *Linear inequalities and related systems,* edited by H. W. Kuhn and A. W. Tucker, Ann. of Math. Studies, no. 38, Princeton Univ. Press, Princeton, N. J., 1956, pp. 183-193; Russian transl., IL, Moscow, 1959. MR **18**, 450.

[7] J. von Neumann, *Zur Theorie der Gesellschaftsspiele,* Math. Ann. **100** (1928), 295-320; English transl., in *Contributions to the theory of games.* Vol. IV, Ann. of Math. Studies, no. 40, Princeton Univ. Press, Princeton, N. J., 1959, pp. 13-42; Russian transl., in *Matrix games,* Fizmatgiz, Moscow, 1961, pp. 173-204. MR **21** #635.

[8] H. Nikaidô, *On von Neumann's minimax theorem,* Pacific J. Math. **4** (1954), 65-72. MR **15**, 816.

[9] G. Š. Rubinšteĭn, *Several examples of dual extremal problems,* Mathematical Programming, "Nauka", Moscow, 1966, pp. 9-39. (Russian) MR **38** #5467.

[10] M. Slater, *Lagrange multipliers revisited,* Cowles Commission Discussion Paper, Math., 1950.

[11] H. Uzawa, "The Kuhn-Tucker theorem in concave programming," Chapter III in K. J. Arrow, L. Hurwicz and H. Uzawa, *Studies in linear and non-linear programming,* Stanford Univ. Press, Stanford, Calif., 1958; Russian transl., IL, Moscow, 1962. MR **21** #7115.

[12] S. I. Zuhovickiĭ and L. I. Avdeeva, *Linear and convex programming,* 2nd ed., "Nauka", Moscow, 1967; English transl. of 1st ed., Saunders, Philadelphia, Pa., 1966. MR **37** #2521; #2523.